Humanity and the Sea

Series Editor
Andrew I.L. Payne
A & B Word Ltd
Halesworth, Suffolk, U.K.

For further volumes:
http://www.springer.com/series/11481

Katsumi Tsukamoto • Mari Kuroki
Editors

Eels and Humans

Springer

Editors
Katsumi Tsukamoto, Ph.D.
Department of Marine Science and Resources
College of Bioresource Sciences
Nihon University
Kameino, Fujisawa
Kanagawa, Japan

Mari Kuroki, Ph.D.
The University Museum
The University of Tokyo
Hongo, Bunkyo-ku
Tokyo, Japan

ISSN 2213-607X ISSN 2213-6088 (electronic)
ISBN 978-4-431-54528-6 ISBN 978-4-431-54529-3 (eBook)
DOI 10.1007/978-4-431-54529-3
Springer Tokyo Heidelberg New York Dordrecht London

Library of Congress Control Number: 2013950089

Printed on acid-free paper

Springer is part of Springer Science+Business Media (www.springer.com)

Preface

Why Are Eels So Fascinating?

A good question—it may be because eels are an important food resource for humankind, or perhaps because they are an interesting model species for scientists studying animal migration and reproductive ecology. There may be another reason, though. Biologically, they are fish, but they appear to be more than that; their unique morphology draws a clear line between them and other fish, and their bodies are not streamlined as are those of ordinary fish, but elongate, closely resembling snakes, which are reptiles. Moreover, eels appear superficially to lack many of the characteristics used to identify fish, for instance, gills, fins and scales, although they do have all these features. Another characteristic of eels that attracts our attention and evokes curiosity is their unpredictable behaviour and inconceivable strength, enabling them to move over wet land and to climb rocks beside high waterfalls. All these features lead to us considering eels as enigmatic creatures or metaphysical entities beyond human intelligence; indeed, they have even been deified in parts of the world.

In this book, the various authors attempt to explain as much as they can about eels, focusing on social and cultural aspects wherever they are found. A wealth of eel-related topics is covered, including eel fishing, resources, distribution, aquaculture, economics, cuisine, environment and ecosystems, idioms, arts, tradition, legends, mythology, archaeology and even memorial services, as shown in Fig. 1.

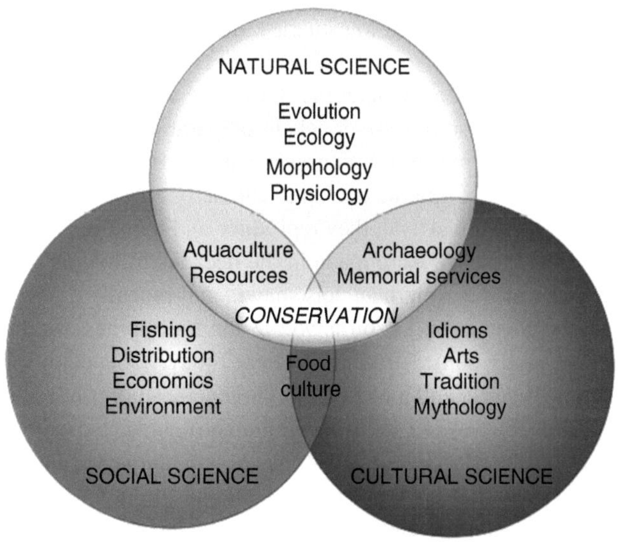

Fig. 1 Various aspects of eel-related science. Comprehensive understanding of such issues enhances public awareness of their importance and helps to conserve the species all around the world

The Natural Science of Eels

Articles or tales about eels often start with or soon mention the word "mysterious". This is because their spawning areas are located far offshore and their spawning ecology was virtually unknown until relatively recent times. Aristotle, the Ancient Greek philosopher and naturalist, famously postulated that eels had to generate naturally from mud, because he could find no larvae or adults with mature gonads. Centuries later, Italian physicians and anatomists took an interest in eel reproduction and ecology. They included Francesco Redi (1626–1697), Marcello Malpighi (1628–1694), Antonio Vallisneri (1661–1730) and Carlo Mondini (1729–1803).

Even so, there was no major breakthrough in knowledge of eel ecology until two Italian scientists, Giovanni Battista Grassi (1854–1925) and Salvatore Calandruccio (1858–1908), made a landmark discovery in 1897. They proved that the fish then known as *Leptocephalus brevirostris*, which resembles a transparent leaf, eventually metamorphoses into the glass eel of *Anguilla anguilla*, the European eel. Until that discovery, the genus *Leptocephalus* had been regarded as a different fish taxon altogether, entirely unrelated to the eel, but with their breakthrough it was finally accepted that freshwater eels develop from the leptocephalus larvae found in the sea. Despite this discovery, it was not until the early twentieth century that superstitions involving autogenesis in eel reproduction—the notion that they originated, for example, from yams (East Asia) or horsetail hair (Europe)—were finally laid to rest by Danish oceanographer Johannes Schmidt (1877–1933). It was he who discovered the spawning area of both European and American eels in the Sargasso Sea. Two of his pupils then continued his work, culminating in further major contributions to the advancement of eel science: Vilhelm Ege (1887–1962) established the taxonomy of freshwater eels in the genus *Anguilla*, and Poul Jespersen (1891–1951) laid the foundations for leptocephalus biology throughout the world.

Thanks to these discoveries, we now know much about the life history of freshwater eels. They are catadromous, spawning in the sea, far from the freshwater they occupy for much of their life. Hatched larvae quickly develop into the larval form known as a leptocephalus, which is morphologically quite different from its appearance as a juvenile (known as a glass eel). Leptocephali grow as they are carried along by ocean currents for long periods after hatching, for anything between 3 months and a year, or perhaps even more. Then, on approaching the freshwater habitats of islands or continents, they metamorphose into glass eels and most enter rivers to swim upstream, when they are known initially as elvers before they settle in a wide range of habitats, from rivers and lakes to estuaries, though some even stay in saline coastal water. Once settled they are referred to as yellow eels, and these grow large on a diet of smaller fish and crustaceans for several years (sometimes tens of years). Then, with the onset of maturation, yellow eels metamorphose into silver eels in preparation for spawning migration back out to sea. The downstream migration commences in autumn and marks the end of the animal's feeding stage, and the silver eels migrate back into and across the ocean to their preferred and specific spawning area where, on arrival, they reproduce and then die.

Various advances in eel science were made in the decades following Schmidt's discovery in 1922 of where European and American eels spawn, and recently a new species (*Anguilla luzonensis*) has been discovered on Luzon Island in the Philippines, the first new discovery since Ege's taxonomic study some 70 years earlier. Today, 19 species and subspecies are described for the genus *Anguilla* and molecular phylogenetic studies have been completed for all the world's anguillids. Surveys of the spawning area of Japanese eels in the Pacific have made huge progress over the past decade alone; newly hatched larvae, spawned eel eggs and adult eels in spawning condition have been collected, and the precise spawning site can be predicted based on oceanographic and geological characteristics of the general spawning area. Further, the behaviour of adult eels during their spawning migration is being studied for European, New Zealand and Japanese eels, using swim tunnels in the laboratory and pop-up tags in the field.

The Social Science of Eels

Eels appeared on earth tens of millions of years ago, but it was only seven million years ago that our own ancestors diverged from apes. We can easily imagine, therefore, that early humans would catch eels in rivers and consume them, starting the relationship between eels and humans. Since then, the relationship between humans and eels has been maintained, largely because of the long-standing human tradition of exploiting eels as food. At first, eels were caught by hands and feet, but as time passed, many different types of fishing gear and methods of catching eels were developed. Eels were caught using a variety of contrivances including spears, scythes, pots, nets, weirs, rock piles, and hook and line. Their unique habits, body form, behaviour and life history gave rise, however, to some common features of fishing gear and methods around the world, and these were put to use in catching glass eels, yellow eels and silver eels, i.e. every stage in the life cycle of eels except for the earliest (eggs and leptocephalus larvae), wherever they were found.

In historical times, all the eels caught from a river would be consumed locally in the village where the people lived, but later, as catches increased, part of the catch not needed for immediate consumption would be preserved and used for trade. This was the beginning of eel marketing and economics. Then, the increasing social demand for eels as food gave rise to eel aquaculture, as a means of maintaining a stable supply of eels to the market. The increased eel consumption in many countries also produced a variety of recipes for eel dishes around the world, including kabayaki, smoked eel, eel pie, jellied eels, angulas, eel stew, eel curry and matelote d'anguille. Cooking knives themselves have been designed specifically for filleting eels according to local needs, and tableware has been invented to handle eel dishes only. These features are all representative of the food culture that has developed around eels.

However, the overexploitation of eel resources and the degradation of many river environments triggered a huge decrease in eel resources worldwide that became notably evident in the 1970s. Now scientists working on eels have to pay close attention to the problems of the decimated resources as well as the need for their conservation. Research aimed at developing the technology to produce glass eels artificially began with European and Japanese eels in the 1960s, with the motivation in both cases being to achieve a stable supply of glass eel seedlings for aquaculture ponds and to reduce the impact of fishing on wild glass eels with a view to their conservation. This research challenge has recently been expanded to other eel species, e.g. the Australian and American species. However, although promising, such efforts have yet to achieve definitive success, meaning the production of large numbers of offspring, in any species of eel.

The Cultural Science of Eels

The traditional exploitation of eels as food spawned a feeling of familiarity with eels in the human mind, and this relationship naturally yielded various cultural references to the species. Eels now appear in idioms, metaphors and proverbs in many languages, e.g. "as slippery as an eel" in English, "il y a anguille sous roche" in French ("there's an eel under the rock," referring to a fishy or dubious situation), and "unagi nobori" in Japanese ("eel climbing," or a phenomenon that continues to rise with no apparent end). In addition, people have expressed eels in various forms of art and daily items, such as poems, paintings, carvings, handicrafts, accessories, books, films, storytelling and fables, children's stories, cartoons, comic songs and even cookies. All of these reflect a human interest and perhaps even affection for eels, proving that this unique fish has always fascinated people.

At the same time as being objects of affection, however, eels are also regarded as objects of fear and respect. This is because they resemble snakes or dragons as well as exhibiting strange behaviour and incredible power. Old sayings, legends and myths about eels survive in various parts of the world, and in these, eels are sometimes portrayed as monsters or gods related to floods and droughts, or as spirits of water. A well-known example is the tragic romance of the eel-god Tuna (who also has other names in mythology, e.g. Tunaroa) and the beautiful village girl Hina. This myth is related widely among the tropical islands of the Pacific, combined with a legend on the origin of the coconut tree (see the similar story outlined in the French chapter 5).

Memorial services are still held for eels in some parts of East Asia, to thank the eels that have been sacrificed by humans and to pray for their souls. Eel mounds or memorial towers have been erected in places where these services are held; solemn, respectful ceremonies to pray for the eels' souls are held in front of them, after which yellow eels are released into the wild. Scientific evaluation of these events has recently started in the hope of finding ways to improve release techniques and thereby enhance eel resources.

Eels are also bound up in taboos or spirituality. The indigenous peoples of Canada and New Zealand treat eels with awe and respect. Further, some people or districts in Japan respect eels as messengers of Buddha, so shun the idea of eating them, as do the folk of some parts of Polynesia and Melanesia where eating eels is considered taboo. In the latter case, this is because people believe that their ancestors eight generations back were eels, so worship them as sacred creatures. This leads to a belief that, if sacred eels are killed or eaten, precious spring water will dry up or people will be afflicted by terrible disease. Many beliefs, totemism and taboos in Pacific islands are linked to eels, because they are important, key animals for local people. It is also of note that the routes by which this culture has been propagated and diffused are closely related to those of human migration and interaction in the South Pacific. At the same time, they overlap with the image of eel migration from spawning areas offshore to freshwater rivers on small tropical islands in the South Pacific, resulting from the dispersal of the tiny leptocephalus larvae in oceanic currents.

Where Does This Lead?

We know that eels have been shrouded in mystery since the earliest days of human thought. At the same time, they are popular in world cuisine and are a familiar part of everyday language and of arts and crafts, revealing a long-lasting and close relationship with humans. In recent decades, however, with global populations of eels in sharp decline, some species face the real threat of extinction, and effective conservation strategies and measures are needed urgently to protect them.

As shown in Fig. 1, this book attempts to explain much of what is known currently about eels, focusing on social and cultural aspects as well as eel science. Lateral comparisons across these issues between various countries should provide an overall image of the relationship between eels and humans, and as such, our ultimate aim is to encourage a more comprehensive and detailed understanding of eels from the perspectives of social, cultural and natural science. By promoting an understanding of this close relationship between eels and humans, we hope to engage the broader public and increase public awareness of the importance of eels, helping hopefully to conserve these unique endangered fish. By so doing, we hope to ensure the survival on earth of these mysterious yet often loveable creatures.

To conclude, we thank Andrew I. L. Payne, Series Editor of Springer's "Humanity and the Sea" book series and Springer itself for giving us the chance to publish this book on eels as the first volume in the series, and for invaluable advice on editing the work.

Kanagawa, Japan Katsumi Tsukamoto
Tokyo, Japan Mari Kuroki

Contents

Eels and People in the United Kingdom

David Righton and Mandy Roberts

The European eel (*Anguilla anguilla*) is arguably the most widespread species of fish in that continent, being found in all countries with a coastline. In the United Kingdom, which has the second longest coastline in Europe, eels are found in almost all rivers and lakes; their ability to leave the water and move across damp ground even allows them to colonize lakes and reservoirs that are not connected to a flowing waterway. Eels are an important part of the freshwater ecosystem, but their abundance has declined drastically across Europe in recent years, as has that of other species of eel worldwide (see other chapters in this book), raising concerns about the state and the health of the stock. Recent research has shown too that the European eel, although commonly described as a freshwater species, also lives in estuaries and the sea. Some of the eels found in those waters have undoubtedly migrated to the sea in preparation for their spawning migration, but there is a proportion of the population in marine and brackish waters that never actually enters freshwater.

The life cycle of eels (Fig. 1.1) is truly remarkable and some of the detail remains a bit of a mystery. Generations of scientists, stretching back to Aristotle, have been both puzzled and inspired by eels, and details of eel biology are continuously being refined. Eels undergo a lifetime of change, beginning at the start of their lives as they hatch from eggs in the Sargasso Sea. The form of those young eels, the leptocephalus larvae (Fig. 1.1) is surprising to many, because they do not look like eels at all, and instead are transparent and leaf-shaped. Their heads are tiny, with ferocious looking teeth, and their bodies, in proportion to their heads and tails, large and gelatinous, acting like sails to help them drift on ocean currents, rather like leaves in the wind. The importance of this shape cannot be overstated, because although leptocephali can swim, their journey of >4,000 km to the coasts of Europe depends critically on ocean currents helping them drift to their destination. This journey seemingly takes between 2 and 3 years, principally in the North Atlantic Drift, a current that crosses the Atlantic Ocean from the eastern edge of the Sargasso Sea to the coastal margins of Europe. During that time, the leptocephali drift in the surface waters of an ocean that is up to 4.5 km deep.

The actual duration of eel larval migration is not known precisely, but they arrive off Europe's western edge in autumn each year. There, they metamorphose into glass eels (Fig. 1.1), tiny, totally transparent replicas of the adults, but no more than a few centimetres long. Once in the coastal margins, they gather in huge shoals and wait for the water to warm, then migrate en masse into estuaries and the lower reaches of rivers. The first glass eels of the season arrive in the estuaries of southern Europe generally in February and March; their arrival in estuaries and rivers farther north and east is slightly later in the year. Some of these eels will remain in the estuaries and complete their life cycle without ever leaving the marine or estuarine environment, but most will progress upstream fairly rapidly and, as they do so, become pigmented and known as elvers. This stage lasts only a few months and, as they grow bigger and more muscular by feeding on the rich invertebrate community of their river or estuary homes, they become more obviously pigmented and are thereafter known as yellow eels (Fig. 1.1).

That stage persists for many years, as many as 90 but more commonly 10–15 years, until they are ready to return to the Sargasso Sea as silver eels (Fig. 1.1), the migratory form of eels that have a single purpose in life, to migrate back to their own birthplace to give rise to the next generation. Remarkably, that phase of the life cycle is very short, thought to be no more than 6 months. In that time, a silver eel will swim at least 4,000 km across the Atlantic Ocean (and possibly up to 8,000 km from the eastern part of the species' range), facing the dangers posed by whales, sharks and other predatory fish that see it as a tasty meal. Remarkably, the silver eel phase is also when the eels undergo sexual

D. Righton (✉) • M. Roberts
Centre for Environment, Fisheries and Aquaculture Science, Pakefield Road, Lowestoft, Suffolk NR33 0HT, UK
e-mail: david.righton@cefas.co.uk

K. Tsukamoto and M. Kuroki (eds.), *Eels and Humans*, Humanity and the Sea,
DOI 10.1007/978-4-431-54529-3_1, © Springer Japan 2014

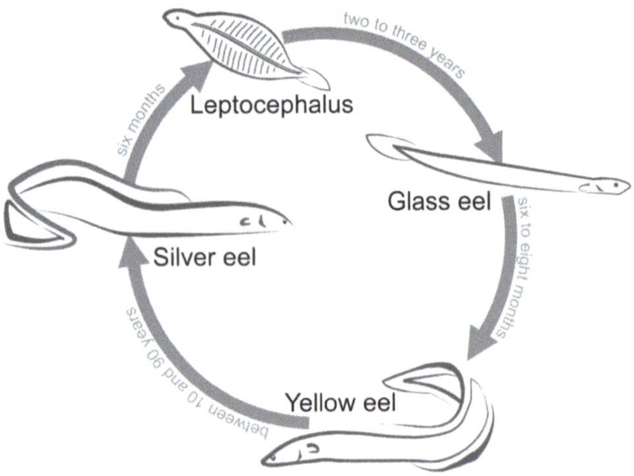

Fig. 1.1 The life cycle of the European eel; unlike most species of fish, eels spawn just once, then die

Fig. 1.2 Image from the Lutrell Psalter, a fourteenth century religious text, showing an eel buck to the right used to catch silver eels upstream of a water mill (at *bottom left*). Source: the British Library

maturation in readiness for spawning. These major events are achieved solely with the fat stores that eels build up during their lives; metamorphosis into the silver stage shrinks the stomach and gut so that the fish can only absorb water. At the end of its journey, the silver eel is probably exhausted, yet it still has to find a mate or mates with which to spawn before its remaining energy is spent. This done, and with a new generation of eels starting life, the purpose of a silver eel's life is fulfilled. There is no return journey.

Eel Fishing

The fact that eels are so widespread in the UK means that they are one of the most widely fished freshwater species in the country. Locals have captured and eaten glass eels, yellow eels and silver eels for millennia, long before society began keeping records. Archaeologists believe that eels were being fished at Lough Neagh in Northern Ireland as long ago as the Bronze Age, eel skeletal remains having been uncovered there when excavating settlements dating back 3,000–4,000 years, and possibly even before that. There is evidence too of eels being fished commercially at Crosskeys, on the Lower Bann Basin north of Toomebridge (near Lough Neagh), from about 2000 BC. Written records of eel fishing extend back to the fourteenth century, as demonstrated by illustrations of eel bucks (a rack of large wicker traps designed to catch migrating silver eels) in the Lutrell Psalter (Fig. 1.2).

Until relatively recently, UK people fished mainly on a subsistence or small-scale basis, so most fisheries simply adopted the traditional methods of their ancestors. Even today, when commercial fishing pressure is intense, many of those targeting eels continue to use traditional methods that vary only slightly from place to place. However, the recent decline of the stocks of eels has resulted in restrictions on fishing activity becoming tighter and more widespread, prompting concern that traditional methods and livelihoods will be lost, perhaps never to be recovered.

Glass Eels

Fisheries for glass eels are highly seasonal because they depend on the annual influx of young eels from the Atlantic. Significant fisheries are documented for the River Severn (SW England), the River Thames (SE England), Lough Neagh and many other waterways. Historical records describe the upriver migration of glass eels as dense ribbons of fish, hundreds deep and tens of miles long. To catch glass eels, all a fisher would have to do was to dip a net into the water and scoop out the glass eels in their tens, hundreds and thousands, day after day during the several weeks of the migration. On the Thames alone, as Cornish wrote in 1902, "they made a black margin to the river, on either side of the banks." Dipnets are traditionally square and regulations demand that a net can only be as large as to permit a single person to use it (Fig. 1.3). Glass eels are still caught in this way, but not in such great numbers, and the number of fishers involved in the fishery has now dwindled, leaving glass eels to be caught mainly to assist eel-stocking schemes within the UK and elsewhere in northern Europe.

Yellow and Silver Eels

As for glass eels, the numbers of yellow and silver eel fishers has dropped notably in recent years. Yellow eels can be caught in a number of ways, using traps or weirs, spears, nets or hooks. The traps vary in size from small baskets made of wicker (using reeds or willow switches; Figs. 1.4 and 1.5) to large-scale eel bucks and weirs. The small traps or nets are baited and typically set from small boats (Fig. 1.4), then checked after a few hours or sometimes days. Types of bait vary, but are typically fish, but in the past the traps might have been baited with meat that had gone bad or, more recently, roadkill, i.e. animals killed on the road! Eels sense the bait and move into the traps, but once inside they cannot escape. The traps themselves are still made the same way as they were more than 3,000 years ago, but larger modern versions made of metal and twine, known as fykenets, are more common these days. Yellow eels can be caught too by baiting hooks, or by tying worms into a woollen ball and then "bobbing" it in a river. Eels strike at the ball, catch their teeth on the fibres of the wool and cannot disentangle themselves. This fishing technique developed because eels are notoriously difficult to unhook from fishing lines. Yet the diversity of fishing methods does not stop there: in places where they used to be common, eels are caught simply by prodding the bed of the river or lake with a multi-tined spear, often termed a gleave. Remarkably, many similar fishing methods are found across the world wherever eels are fished, a case of convergent technical evolution!

Fig. 1.3 Fishing for glass eels in the River Parrett, Somerset (photo Richard Dearnley, Environment Agency, UK)

By contrast, the capture of silver eels relies on their downstream migratory urge, and generally requires the fisher to channel the eels into a trap or net of ever diminishing diameter, or to force the eels to swim through barriers on weirs through which they cannot then return, allowing them to be picked out; they are often then stored in barrels or pots. As they tend to be fixed in once place, many of the traps used to catch silver eels are quite magnificent feats of engineering that have existed for many years. Some, like the wicker eel bucks that can be traced back to at least the fourteenth century and which were used until the early twentieth century (Fig. 1.6, but see Fig. 1.13) can be raised from the water at

Fig. 1.4 An eel "grig" or trap being deployed from a small vessel. These traps were in common use in historical times (reproduced with the permission of the Oxford History Centre)

Fig. 1.5 Peter Carter (*left*), the last remaining "eel man" of the East Anglian fens, builds an eel trap. The traps that Peter weaves are no different from those used by his recent ancestors (*right*), or even those that lived more than 2,000 years ago near his present home. Photograph reproduced with the permission of Eastern Daily Press, Norwich

Fig. 1.6 An illustration (*top left*) of eel bucks, large wicker baskets that can be lowered into a flowing river to catch silver eels. Examples of these are documented in nineteenth century photographs at Harleyford (Buckinghamshire, *top right*), Caversham (*bottom left*) and Bray (Berkshire, *bottom right*). Images reproduced with the permission of the Oxford History Centre and English Heritage

Fig. 1.7 Eel racks take advantage of the river's flow and the downstream migration of silver eels, as shown in this photo (*left*) of an eel rack on the River Avon taken in the 1960s. More recently, Roger Castle, one of the UK's last remaining silver eel fishers, surveys the run of silver eels emerging from his eel rack (*right*). In the past, a monthly run of eels might have exceeded 3 t, but today, catches are closer to just 10 % of that figure (photos provided by Roger Castle)

the fisher's whim to undertake maintenance and to harvest the eels trapped inside. Others channel the eels to the water surface to be harvested; an example of such a device is the eel weir on the River Stour in southern England that is still used by local fisher Roger Castle today (Fig. 1.7). In a few cases, entire buildings were built to channel and harvest silver eels, an example being the eel house at Alresford that is being restored currently to its former glory as public interest in eels and their historical significance burgeons. Unlike yellow eels, silver eels are fished in autumn only, because that is when the eels head downstream (i.e. run) on their way to the Sargasso Sea. Catches tend to peak when rainfall is heavy enough to produce fast and voluminous river flow and the moon is new—conditions that help the silver eels travel quickly and undetected. As have the catches of glass eels, so too have catches of yellow and silver eels dwindled during the past few decades, and concomitantly, the number of local fishers catching them has declined sharply.

Eels in Human History

In medieval times, eels were sometimes used as a form of currency as well as a source of food. There is evidence, for instance, that they were used as one-off payments or as an annual payment for rent of land or services from the lord of the borough, or the priests of the local abbey or diocese. An excellent example of this can be found in the city of Ely (pronounced "eel-y," so named because of the importance of the local wetlands, known as the fens, for catching eels). Archaeological excavation there has revealed wicker eel traps 3,000 years old that are more or less identical to those used in the area today and which, in historical times, would have been used to catch the eels used as payment to the local parish of Ely. Records show that the Lord of the Manor there was paid by his serfs an annual rent of 100,000 eels. One fisher of this lineage, Peter Carter, is alive today (Fig. 1.5), and recalls that 80,000 eels had to be paid to the bishop of Ely annually, and another 100,000 to the bishop of nearby Ramsey. A prominent reminder of the extent of these payments still exists in the form of Ely cathedral (Fig. 1.8), which was built of stone paid for by eels, and the fenland tradition of a wedding ring made of eel skin has not yet disappeared completely, although it is not very commonly seen today! Indeed, the residents of Ely still gather each year to celebrate the eel in the annual eel festival (Fig. 1.8), and the historical importance of eels to Ely is recognized in art exhibits and an eel trail.

One of the best-documented examples of early commercial eel fisheries are those of the Rivers Thames and Trent, both of which are documented in the Domesday Book (Anon 1086). That book also lists hundreds of watermills across England whose rent was paid in eels (referred to as "sticks," but actually consisting of between 20 and 25 eels). The trade in eels between the fisheries of London and other nations is described in considerable detail in the records of the port of London, and drawings of the time helped to reflect this fact (Fig. 1.9). Eels, being easily transported and kept alive in even the most basic conditions, were a valuable commodity, and in London were vital to that city's prosperity, and to its

Fig. 1.8 Ely cathedral, which was built of stone paid for in eels (*left*: source Wikipedia, Creative Commons License), and a procession celebrating the eel-y history of Ely (*right*: photo provided by Chris Hunt, Ely People)

Fig. 1.9 An eel seller advertising her wares on the streets of London, painted by Francis Wheatley in his *Cries of London* series, produced between 1792 and 1795 (Heritage Images)

citizens' stomachs, particularly in times of hardship. The trade in eels in the UK was actually quite notable until the early part of the twentieth century, but the economic importance of eels has long since declined.

Remedies and Superstition

Aside from their use as food or in trade, eels have been considered to hold properties that can help people in other ways. One medicinal application was described by Culpeper (1653): "Eels, being put into wine or beer, and suffered to die in it, he that drinks it will never endure that sort of liquor again." This is not an isolated suggestion, and it is replicated quite widely in the literature. Others are perhaps no more than ritual uses: warts were said to be cured by rubbing eel blood on them, deafness cured by dropping the oil from an eel's liver into the ear canal, and wearing a garter of eel skin is said to cure cramp or arthritis. Similar cures can be found in the literature across Europe, suggesting that such remedies were widespread and, perhaps, had at least some basis in fact.

Non-medicinal uses are equally curious and useful, but those mentioned here are confined to the UK. For example, it was said that wearing an eel skin jacket would protect witches and sorcerers from bullets. More amazingly, it was thought that eel fat would render fairies visible to humans (it is not clear from the literature whether the eel fat has to be rubbed on the fairies or eaten by those looking for them!). For those interested in looking into the future, it has been recommended to eat an eel heart to provide the power of prophecy. However, eating too many eels was apparently to be avoided because, according to folklore, eating a whole eel would strike one dumb!

As with many other countries in Europe, the eel has become associated with people who are not trustworthy, or are evasive. The phrase "slippery as an eel" is often used for such people, where the word slippery is not literal, but instead is used generally as a metaphor for dishonesty. Curiously, the fenlands in the east of the UK appear to have generated a fairly unique metaphor, but this time with no apparent meaning (other than a bit of fun): those born in the fens have been described as "yellow bellies," like their eels, dating back to Grose's (1787) "A Provincial Glossary."

Legends and Monsters

The mysterious nature of eels provides rich material for stories about the origin of eels as well as their relationship to the origin of UK peoples and culture. For example, a common belief was that a long black horsehair thrown into a running stream instantly becomes a live eel or water snake. Correspondence in Notes and Queries (1886–1887) under the heading "Animated Horsehairs" indicates that this had been a widely held notion in England, Scotland and elsewhere, at all levels of society well into the late nineteenth century. In the journal *Philosophical Transactions of the Royal Society* (Fig. 1.10), Dale (1698) wrote "Many there are, that with Aristotle, will have the generation of eels to be spontaneous or equivocal, and will not allow the distinction of sex; the difficulty of how eels should come to be in any pool, pond, moat or ditch, in which never any were put, they are not able to surmount, and therefore have vainly imagined them either to be produced from mud or from a peculiar sort of dew, falling in May or June, upon the blades of the grass, whereof turfs being cut, and the grassy sides being together, and then laid on the warmest side of a promising pool or pond, the sun's heat will then hatch them." Even today, the origin of eels remains something of a mystery!

Large eels have sometimes been the basis of stories about monsters or mythical creatures. For example, there is, or was, a tradition in Ireland and Scotland that monstrously large and ugly "hairy eels" existed. The Irish called these creatures horse-eels (or kelpies), supposedly because their heads and foreparts resembled those of horses, while their tails were those of eels. Such tales can shed light on other monster stories, the most famous modern example likely being that of the Loch Ness monster, sightings of which may have been based in part upon glimpses of large eels at the water's surface.

If not the origin of the stories about "Nessie," as the Loch Ness monster has come to be known, large eels provide a fascination for many, and the search for monstrous eels is not confined to the history books. The UK has a strong tradition of recreational angling (angling for fun, competition or relaxation), first documented by Izaac Walton in The Compleat Angler (or The Contemplative Mans Recreation), published as Walton (1653). Today, the UK's National Anguilla Club (NAC) is a thriving institution whose near 100-strong membership is always striving to catch the biggest eel possible (Fig. 1.11). The current record stands at 11 lb 2 oz (~5 kg), but the variety and length of the specimens caught by the NAC over the past 25 years is impressive,

Many there are, that with *Ariſtotle*, will have the Generation of *Eels* to be *Spontaneous* or *Equivocal*, and will not allow them the Diſtinction of Sex; the Difficulty how *Eels* ſhould come to be in any Pool, Pond, Mote or Ditch, in which never any were put, they are not able to ſurmount, and therefore have vainly imagined them either to be produced from Mud, or from a peculiar Sort of *Dew*, falling in *May* or *June*, upon the Blades of the Graſs, whereof Turfs being cut, and the Graſſy Sides being together, and then laid on the warmeſt Side of a well promiſing Pool or Pond, the Sun's Heat will thence hatch them; of which Original *Myllius*, in his Treatiſe *de Origine Animalium*, *lib.* 10. and *Morhoſius de Metallorum tranſmutatione*, *p.* 38, 39. ſeem ſo well ſatisfied, that they give the Proceſs of this Affair, as practiſed by the *Dutch*, ſucceſsfully to ſtock their Fiſh-ponds with that ſort of Fiſh.

Fig. 1.10 The original text from the Philosophical Transactions of the Royal Society in 1698, detailing Mr P. Dale's notes about how eels came to be

Fig. 1.11 The impressive British record eel, or "Terry's eel" which weighed in at 11 lb 2 oz (~5 kg: *centre*). Photograph reproduced with the permission of the National Anguilla Club

though perhaps not as impressive as the dedication of these anguillophile anglers.

Eel Cuisine

The importance and abundance of eels in the UK in historical times was not just associated with trade. Eels were an important food and source of protein, partly because fresh eel meat could be obtained readily, but also because eel meat could be easily preserved in salt or, more commonly, by smoking. Recipes for eel are numerous, but particularly popular were stews and casseroles, in which eel meat is cooked slowly for a long period, becoming tender and sweet.

Fig. 1.12 Jellied eels (*top left*) can still be found in modern supermarkets, although they are less popular than they used to be (*bottom right*). The tradition of jellied eels was for many years popular in London, as shown in the photograph (*top right*, Corbis Images) taken in the 1950s. "Eel and Pie" shops still exist today (*bottom left*, source: Wikipedia, Creative Commons License)

An eel dish that is, perhaps, most strongly associated with the UK is jellied eels (Fig. 1.12), produce found nowhere else in the world. Jellied eels first became popular at the end of the nineteenth century, and quickly became the Cockney (i.e. a Londoner, but in reality a person from the east end of that city) national dish (Fig. 1.12). For some, the appearance and taste of the eels can be offputting, although it is not hard to understand why. The recipe for jellied eels is simple: chop the eel, boil it in stock and allow the mixture to set. Once cold, jellied eels are eaten on their own or, more traditionally, with mashed potato. Jellied eel can also be sealed into a pastry crust and served with mash, and typically with a green sauce made from parsley and termed "liquor." In the early days of the UK's "Pie and Mash" or "Eel and Pie" shops (Fig. 1.12), which peaked during the reign of Queen Victoria in the nineteenth century, it is said that unscrupulous merchants would use almost anything in their pies, including rotten eels, and disguise the contents using copious quantities of pepper. Perhaps unsurprisingly, this practice sometimes led to outbreaks of food poisoning! Pie and mash shops are still found throughout London even today, although eel pie provides less and less of their business. Jellied eels are certainly less popular than in their heyday of the early twentieth century through to the 1960s, but it is still possible to find them in supermarkets across the country.

The strong association between eels and London probably dates back to the time of the Great Fire of London in 1666, when the capital was ravaged by fire and, after which, many Londoners faced starvation. To help prevent a catastrophe, live eels in vast quantities were brought up the Thames by Dutch eel barges known as "eel schuyts," helping to save many lives. In consequence, a Royal Decree was awarded to the Dutch by King Charles II, allowing eel barges to moor gratis on the Thames (provided that the mooring was never left vacant for more than 2 min!). The barges were also given exclusive rights to sell eels, and between them sold up to ten million eels a year. This Decree was in place until 1938, when the last remaining barges packed up and left in response to declining trade. Today gourmet restaurants and smoke houses offer fillets of smoked eel to customers, praising their rich, sweet, firm flesh, high in vitamins and proteins. This helps to keep eels within the nation's consciousness. Perhaps ironically, the decline in interest in eels has left relatively few establishments that prepare and sell the product, so the products of smokeries and fish delicatessens have now become somewhat niche and sought-after.

Fig. 1.13 Father William balancing an eel on his nose in the illustration from the poem in the book *Alice in Wonderland*. Image drawn by John Tenniel, taken from Lewis Carroll's book *Alice in Wonderland*. A weir with eel bucks is clearly depicted in the background (Cutcaster Images)

Eels in Literature and Art

Eels have not escaped the strong tradition in English literature, and have been mentioned in many texts. It is fitting that the most famous English writer of all, William Shakespeare, wrote of eels in the fifteenth century in King Lear. There are doubtless almost countless other examples in literature of the role eels played in general and everyday life, and many of these are accounted for in the excellent book of Fort (2002): "The Book of Eels." Yet descriptions of eels are not always mundane. One example is provided in the children's fantasy book "Alice in Wonderland" (Carroll 1865). This English classic contains a poem that describes a conversation between an idiotic youth and the venerable Father William. The section referring to an eel, and accompanied by a wonderful drawing (Fig. 1.13; note the eel bucks in the background) is as follows:

> You are old, said the youth, one would hardly suppose
> That your eye was as steady as ever;
> Yet you balanced an eel on the end of your nose –
> What made you so awfully clever?

Eels in the visual arts are less common, perhaps because the subject matter is not considered sufficiently attractive, or maybe because eels do not inspire anything more than functional works of art. However, even these can be rendered beautifully. Houghton's (1879) illustrations in his book of freshwater fish bring yellow eels vividly to life, as do Robertson's (1875) images of eels on shore, which also serve as a reminder that eels can tolerate being out of the water for some time if necessary (Fig. 1.14). Bristow's depiction of eel fishing painted in 1812 suggests a gentle and interesting life as an eel fisher (Fig. 1.15), as does Myles Birket Foster's water colour painting of eel bucks dating from 1890.

Much earlier than these formal paintings, eel fishing features in a well known medieval text, the Lutrell Psalter. Put together by a scribe and his artists at a monastery near the fenlands of Lincoln, a centre of eel fishing, in the mid-fourteenth century, one page (Fig. 1.2) shows a watermill and, upstream, a set of eel bucks that would have been used to catch silver eels. Returning to the present day, we recently came across a drawing of the eel's journey to the Sargasso Sea (see the Epilogue) that depicts a benevolent eel carrying an exhausted couple and their child across a vast ocean. The artist, Peter Gander (a resident of Whitstable in Essex, a place once replete with eels), says this evocative image was inspired by a childhood dream.

Eels and Science

The science of eels in the UK has a long history, and there are many famous UK eel scientists, past and present, that are too numerous to mention here. One of the most famous, or perhaps notorious, is a former employee of the UK's natural history museum, Denis Tucker. Convinced that European eels were the same species as American eels, Tucker wrote and published a paper in the prestigious journal *Nature*, under the title "The Atlantic eel problem" (Tucker 1959). In that paper, he claimed that eels in Europe were, to simplify the argument, American eels that had drifted to Europe and which had no chance of returning to the Sargasso Sea to spawn. He provided a number of elegant arguments in support of his claims, but his work was not well received. The Archbishop of Canterbury even became involved in the controversy about the truth of the work! Tucker was a sometimes difficult man, and he refused to back down or moderate his claims. This led to his eventual

Fig. 1.14 Robertson's pencil sketch of eels on the shore of the Thames (reproduced with the permission of the Oxford History Centre)

Fig. 1.15 Edmund Bristow's 1812 "The Eel Catcher," reproduced with the permission of the Ferens Art Gallery, Hull Museums

dismissal from his job at the museum and, sadly, he was never able to find another scientific position. He died in summer 2009, destitute and discredited. However, later that same year, we received a mysterious and salient e-mail from "the ghost of Denis Tucker" in response to the publication of a short paper that the first author of this chapter co-authored on the migration of eels to the Sargasso Sea. Perhaps Denis still has an influence on the science of eels!

Our work at Cefas, and more widely in Europe, is now gaining the attention of the public through direct and indirect methods. Since 2008, we have been implanting electronic recording devices (tags) into migrating silver eels. The tags are made to be positively buoyant by attaching small floats to them, which makes them look a little eel-like! When the eel dies, the tags are released and float up to the sea's surface, where they then drift at the mercy of currents. Some of these tags wash up on coastlines and some are discovered by beachcombers or holidaymakers (Fig. 1.16), who then return the tags to us. It is usually the case that the people who find the devices are highly excited by the "treasure" they have

Fig. 1.16 A modern-day message in a bottle, designed to fit into an eel and record details of its spawning migration (*top left*). The recovery of tags depends on the cooperation of an increasingly interested public, as shown *top right* (photo Rob Brookes) and bottom (photo courtesy Tycho Anker-Nilssen, Norwegian Institute for Nature Research)

found, and many newspaper and internet articles have been published on the discovery of eel tags. The data that the tags have been collecting provides information that cannot be collected any other way, and provide a fascinating window into the oceanic lives of eels. By telling people who have found tags about the data that their tag has collected, we have found that they become even more enthusiastic about eels, and usually they want to help spread the word about eels and why they are a valuable species that needs careful management to conserve them for future generations. As scientists, our contribution to the conservation of eels is to gain new knowledge that will help others make good decisions about habitats and livelihoods, and it is rare that the wider public can get involved. Therefore, through our tagging work, we have been able to involve a few members of the public, and we and they are helping to spread the message about how important and interesting eels are and can be. Our hope is that this scientific work will continue to engage the public, young and old, and that by increasing public awareness of the importance of eels and generating interest, the European eel will not become a forgotten fish.

References

Anon (1086) The Domesday book. In: Morris J (ed) English translation 1977. Phillimore, London, Unpaginated

Carroll L (1865) Alice's adventures in wonderland. Macmillan, London, 192 pp

Culpeper N (1653) The complete herbal. Reprinted in 1863 by Thomas Kelly, London, 398 pp

Dale P (1698) Some considerations about the generation of eels. Philos Trans R Soc 20:90–97

Fort T (2002) The book of eels. Harper-Collins, London, 288 pp

Grose F (1787) A provincial glossary: with a collection of local proverbs, and popular superstitions. Hooper, London, Unpaginated

Houghton W (1879) British fresh-water fishes. William Mackenzie, London, 204 pp

Notes and Queries (1886–1887) s7–II, animated horsehairs. Oxford University Press, London, Various pages

Robertson HR (1875) Life on the Upper Thames. Virtue, Spalding and Company, London, 214 pp

Tucker D (1959) A new solution to the Atlantic eel problem. Nature 183:495–501

Walton I (1653) The compleat angler, Printed by T. Maxes for Rich. Marriot in St Dunstan's Churchyard, Fleet Street, London, 148 pp

Eels and People in Ireland: From Mythology to International Eel Stock Conservation

2

T. Kieran McCarthy

Pleistocene glaciations left their mark on many aspects of Ireland's flora and fauna. The biodiversity of the island's freshwater fauna reflects the incomplete post-glacial recolonization by many taxonomic groups following their total displacement during the time Arctic and boreal conditions prevailed. The absence of many species of freshwater fish, widespread in the nearby landmasses of Great Britain and northwestern Europe, has often been noted in biogeographic discussions on Ireland's fish. Giraldus Cambrensus, a British monk who was chaplain to English King Henry II, came to Ireland with the invading Norman army in the twelfth century and commented on this aspect of Ireland's natural history, chronicling the events he witnessed then (O'Meara 1951).

The indigenous fish species of Irish inland waters are typically euryhaline migratory forms or landlocked populations of migratory species. The catadromous European eel *Anguilla anguilla* is one of the most common and widespread fish of Irish lakes and rivers, and lives also in marine, littoral and estuarine habitats; it was recorded in 33 of 38 coastal lagoons surveyed by Healy (2003). Moreover, studies on otolith microchemistry (Arai et al. 2006) and stable isotope ratios (Harrod et al. 2005) have revealed the life-history flexibility of eels and their facultative catadromy. The main Irish populations of eels are in larger mesotrophic/eutrophic lowland lakes and rivers, where exploitation has taken place for about at least eight millennia. The changing environmental conditions that allowed faunal colonization to proceed about 10,000 years before present also provided opportunity for the human settlement that marked the start of Mesolithic communities in Ireland. The broad current geographic range of the European eel, from Norway to North Africa, suggests that eels might have extended their range north as the ice masses retreated, if oceanic circulation patterns favoured such dispersal by their oceanic larvae. New information on the spatiotemporal evolution of the Irish Ice Sheet and of the effects of changes in relative sea levels allows coastline dynamics to be modelled (Edwards and Brooks 2006). Palaeogeographic maps can now be drawn, permitting better understanding of the early colonization of Ireland's inland waters by fish of marine origin, such as eels.

The apparently reciprocal distribution of European eels and the more-marine conger eel (*Conger conger*) might be explained when ecological studies of both in adjacent estuarine habitats progress. Indeed, studies on the parasites of Irish eels provide insights into their biogeographic status (McCarthy et al. 2009). Like other indigenous fish host species, eels in Ireland are infested by more parasite species than more-recently introduced fish. Irish eel parasite communities include host-specific specialists and generalist forms that are typically noted in recent additions to the island's freshwater fish fauna. Adverse effects of species introductions include those of the invasive swimbladder parasite *Anguillicoloides crassus* that was brought accidentally from Asia to Europe.

Eels and Irish Place Names (*Dinnsheanchas*)

Most place names (*logainmneacha*) used today in Ireland are anglicized versions of old Irish language (*Gaeilge*) names, though some Viking, Old English or Norman names have survived to modern times. Some even older names, such as those of some mountains and rivers, appear to have been adapted from some used by non-Gaelic speaking predecessors. Many place names refer to features of the physical landscape, tribal occupants of a territory, personal names, historical battles or man-made structures, such as fortifications or monastic settlements. Place names have regularly been celebrated in mythological tales, songs and poetry. With the establishment of the cartographic Ordnance Survey in Ireland in 1824, there was a comprehensive systematic naming of Irish places, and many older names were adapted

T.K. McCarthy (✉)
School of Natural Sciences and Ryan Institute,
National University of Ireland, Galway, Ireland
e-mail: tk.mccarthy@nuigalway.ie

K. Tsukamoto and M. Kuroki (eds.), *Eels and Humans*, Humanity and the Sea,
DOI 10.1007/978-4-431-54529-3_2, © Springer Japan 2014

to standard anglicized forms. This provided nineteenth century scholars with a wealth of information. Toponymic studies continue to fascinate modern researchers and local historians (Ó Murchada and Murray 2000; Flanagan and Flanagan 2004). An official body (*An Coimisiún Logainmneacha*) and specialized units in higher education institutions provide researchers and the general public access to extensive databases and expert knowledge on the subject.

In general, animal names are not commonly incorporated in place names in Ireland. The Gaelic word for eels (*Eascann*) is, though, represented in some place names, including Carricknanaskin (*Carraig na nEascann*; meaning the rock of the eels), Cloneska (*Cluain Eascann*; eel meadow), Annaghaskin (*Eanach Eascann*; eel marsh), Loughanaskin (*Loch na nEascann*; the lake of eels), and Trasruhannanaskin (*Trá Shruthán na nEascann*; the strand of the eel stream). There may be other places too with names associated with eels and eel fishing, but no systematic study of this topic has been undertaken yet. Nineteenth century cartographers recorded many eel weirs on their maps, and owners of restaurants (e.g. "The Silver Eel" near the River Shannon, Co. Roscommon) and holiday cottages have more recently made use of the word eel in the names of their buildings. A bridge on a Co. Cork road between Mallow and Buttevant, for instance, was known locally in the 1920s as "The Eel Weir Bridge."

The Gaelic word for a weir is *"cora"* and this is a component of names of many sites that formerly had fishing weirs nearby. Although many such locations appear well-suited to eel fishing, the types of fish captured are not indicated in the place names and in many cases the primary target of fishers may have been salmon or other indigenous species. Such place names may refer also to localities near or at estuaries, as well as sites located well inland. A simple example of such a name is Corry, which is in Co. Leitrim. However, the name form is generally more complex, as in Corrofin (*Cora Finne*), in Co. Clare, which has been variously interpreted as "Finn's weir" or the "Weir of the white water." Other widespread place names that refer to weirs include those that indicate a townland associated with weirs (*Baile na Cora*) which may be in the form Ballinacurra, as in Co. Cork, or Corbally, as in Co. Cork, Co. Limerick and elsewhere. Likewise, the presence of a fishing weir is suggested by place names such as Drumcar (*Droim Chora*; meaning the ridge of the weir), in Co. Leitrim. Association of an eel weir with its owner's residence may be indicated by the names of fortified buildings such as Castlecor, the "Castle of the weir," on the River Blackwater, Co. Cork, or Carrigacurra Castle, the "Castle at the rock of the weir" in the upper River Lee in Co. Cork (Flanagan and Flanagan 2004). The existence of an ancient eel-fishing weir there is mentioned by Smith (1774) in his History of the County and City of Cork.

In some cases, the old name has been replaced by an entirely different English one, as for example in Co. Cork where the town of Midleton is still called *Mainistir na Corann* (the monastery of the weir) in Gaelic. Presumably, the establishment of a Cistercian abbey by Anglo-Norman monks in the twelfth century led to the acquisition of fishing rights by the monastic community and the developing town took its name from the religious settlement. The River Owenacurra, which flows through the town, takes its name from former fishing (*Abhann na Cora*; the river of the weirs) activities. Another example of an Irish town with an English name that is not derived from its original Gaelic one is Newmarket on Fergus, in Co. Clare, which was formerly known as *Cora Chaitlinn* (translated as Cathaleen's weir). According to local historians, an eel weir on the small river that flows nearby was an important source of food during Ireland's devastating famine of the 1840s.

A place name familiar to many Irish people, though rarely visited nowadays, is Kincora (*Cinn Coradh*; the head of the weir) and this refers to a historical site a short distance upriver from the well-known River Shannon eel fishing weir at Killaloe. The site is known because of its association with Brian Boru, a High King of Ireland in the eleventh century. The long history of eel fishing in that section of the river illustrates how names associated with the word weir (*cora*) can be based on that fishing activity.

A place name in the inner part of Cork harbour may also derive from a term used for a fishing weir. The term "kettle" is an archaic Anglo-Norman word for a fish trap still used in English phraseology ("a different kettle of fish") to imply a different set of circumstances. Whereas the name Dunkettle, a short distance from the city of Cork, has been generally assumed to mean the "fort of *Ciotal*" (referring to a personal name), its earliest spelling (Dunkytill in 1301) may suggest a Viking or Anglo-Norman source.

Eels and Irish Folklore

The eel has a generally negative image in Irish folklore, perhaps related to its superficial resemblance to a snake. Ireland has no snakes, though, and this was popularly attributed to a story about St Patrick who was believed to have banished them from the island in the fifth century. It seems that giant eels are among the most common of the various types of lake monster described as inhabiting Irish lakes and that they may be said to exist still in the collective Irish imagination (Danagher 1964; Dunne 2009).

Some of the many wells associated with religious events and local festivals throughout Ireland may have been regarded as sacred in pre-Christian Ireland. The Festival of Lughnasa was celebrated from prehistorical times on mountain tops, lakeshores and at other sites at the end of July or beginning of August and seems to have been an important festival throughout the Celtic world. The subject was comprehensively

Fig. 2.1 A modern print of a lake monster being driven by St Finnbar from Lough Allua in the River Lee by Keith Kennedy (private collection)

reviewed by MacNeill (1962) and, in dealing with assemblies at wells, she described the dread that local people had of Poll Taighe, a place in Co. Leitrim where a mountain stream disappeared deep into a limestone fissure where they believed a horrible "worm" lived. The recurring term "Wormhole," and the more specific term "Eel Hole" in the Diocese of Clogher suggest association with large eel-like monsters. Such demons, often in the form of eel- or serpent-like creatures, featured in many folktales and were often said to live in local lakes. "Lough Sunday" was the term given to the celebrations that took place at Lough Owel in the River Shannon catchment area, where a multitude of people assembled annually even in the nineteenth century. According to local accounts noted by MacNeill (1962), a monster was said to have caused the drowning of a man called Peter O'Donaghue, who in 1818 was indulging in the ancient custom of swimming his horse in the lake during the festival. Nowadays, the lake is known for its very large eels, but folklore and belief in the monster is largely forgotten. In many instances, a lake monster ("*péist*" or "*píast*") is confronted in folktales by a saint, who kills or confines the creature to the depths of a lake (Danagher 1964). A well-described example relates how Saint *Cúán* tricked a much feared *píast* into going to the bottom of Loch Chráilí with a cauldron on her head till "Lá an Luan," which the saint meant as "The Monday of Judgement Day" rather than the more usual "Monday" interpretation. Therefore, it is said that the angry creature is still confined to the lake bed (MacNeill 1962). St Finnbar, patron saint and founder of the City of Cork, spent many years at a remote hermitage in Gougane Barra Lake in the upper reaches of the River Lee, and local folklore credits him with having driven a *píast* downriver and out to sea (Fig. 2.1). St Ciarán of Clonmacnoise is said to have, with book and candle, driven away a monster that infested the valley of the River Shannon in the sixth century and that he banished it forever to the depths of Lough Ree. However, even as recently as 1960, local newspapers reported claims by anglers that the giant 6-m-long serpentine lake monster had been seen again (Dunne 2009). Sometimes, lake monsters were described as being of the waterhorse type, but the eel form was also prevalent and in Connemara there are well-publicized nineteenth century accounts of giant eels being seen, though notably not captured. In small lakes near Clifden in Co. Galway, a local man reported seeing a "horse-eel" in 1961, and this prompted visits by TV crews and zealous monster-hunters/cryptozoologists. However, despite much wishful observation of the lakes, and reference to historical accounts of horse-eels and other lake monsters, there were no further sightings (Dunne 2009). A recent Irish language TV documentary dealing with the Connemara horse-eels kept the stories alive for another generation.

Frequently, interesting folklore and superstitions have been recorded concerning special "holy wells," and quite a few of them are said to have been inhabited by "supernatural" fish. Sight of the fish was often taken to indicate that a pilgrim's wish would be granted. Often the fish is described as being a trout or a salmon. In some cases, however, an eel is said to inhabit the well. Near Dingle in Co. Kerry, a holy well called Tobermanaghan (*Tobar Mancháin*) is said to be inhabited by a special eel. Likewise, near Walshestown in Co. Cork, there is a holy well known as Sunday's Well where a remarkable eel is said to be seen occasionally. It has been claimed in the past that water taken from some wells inhabited by such eels could not be boiled (Logan 1980). St Fanahan's Well, named after the patron saint of Mitchelstown in Co. Cork, is now marked by a sculpture showing the warrior monk with his battle staff ("*Cinn Cathach*"). The statue has a carved eel that reminds pilgrims of the possibility that they might see an eel, a sacred fish, in the well. Local folklore suggests that sight of an eel in the well will result in a prayer being answered and that this would be especially true if the eel swam in a cross-pattern. Although pilgrims visit the well in much smaller numbers now than in former times, a special feast day is still recognized locally on 25 November. The well is said to be especially associated with cures for diseases of eyes, throat and forehead. St Sylvester's Well, at Malahide in Co. Dublin, was also thought to provide cures for ailments and up until at least the 1890s, an eel was placed annually in the holy well.

The records of the Irish Folklore Commission have many references to published and unpublished information collected in the early twentieth century by folklorists talking to local people in remote rural communities. Among the accounts recorded are many about remarkably large eels and lake monsters. There are also many accounts of folklore associated with freshwater and conger eels. Some tell about cures possible using eels. For example, it was reported that a cure for a swollen and sore wrist could be obtained by binding the wrist with an eel skin. Lough Neagh eel fishers still use eel skin in this way (Donnelly 1986). Another folklore account stated that a bald person should "Put an eel in a bottle. Cork it tight and bury it to a dung-hill. Let it there for a fortnight and when you take it up the eel will be turned into oil, an awful lot of oil. Rub that into your scalp and your hair will grow again as good as ever." This liquid was also recommended for treatment of injuries to horses that had cut their knees by falling. To cure deafness, it was suggested that a few eels should be put in cabbage leaves under the fire until they softened, then to squeeze the juice out of them and put that in the ear to cure the deafness. A local Donaghue family had, according to a contributor from Co. Leitrim, received their witchcraft and healing powers from eels. He claimed that "At certain times of the year the eel used to visit the Donaghue houses. It always came flying in the air." It was believed that an ancestor, the original Donaghue, would always return an eel when fishing and that the reputation his descendants had for folk medicine and as healers derived from that habit. Eel fishers are also a source of many stories about eels. Went (1945) noted that River Erne eel fishers liked to fry eels and that they carefully retained the oil and fat from the pan and bottled it, because it was in great demand as a cure for rheumatism. A folk cure for ear-ache described by Wilson (1943) involved heating an eel in a jar placed in a pot of boiling water so that its oils and fat could be obtained. This oily material was then placed in a diseased ear.

In Ireland the traditional lifestyle of travelling communities, as in Roma and Gypsy communities elsewhere in Europe, is undergoing rapid change. Many Irish Traveller families are settling in urban homes and caravan-halting sites, and no longer follow the rural roaming habits of their ancestors. However, unusual examples of their folklore remain to be fully documented. This is illustrated by a recent event observed by eel researchers monitoring a fishing crew involved in eel conservation fishing at Portora near Enniskillen, on the River Erne. A Traveller, who had approached with his wife and infant child from Co. Longford in the Irish Midlands, came to the fisher and asked for an eel. The fisher gave him one and the Traveller proceeded to his vehicle, where he touched the eel's head to the child's tongue. He then took the eel to the river, released it and observed it swimming away downstream. He explained that the child was inclined to protrude her tongue excessively and that this was a traditional cure. It is not known if the cure was successful, because the travelling family left and made no further contact with the fisher. However, someone familiar with the family of the Traveller's wife commented at the scene that he thought this habit was a family trait and that no medical treatment, whether with conventional medicine or folk cures, could affect a change in the child's behaviour. This curious example of folklore involving eels may reflect a wider interest in eels among travelling communities in the past than was the case in the general Irish population, and it merits further study.

Irish folklore records also include stories about eels that came back to life, sometimes sought revenge, and which made their way back to the river or sea. For instance, there are accounts of eels having been seen travelling with pots on their heads or rolling in hoop form away from the house where they were being cooked. Sometimes they are said to have entered houses and eaten food. One was said to have eaten four chickens but that these were recovered alive from the inside of the eel when it was killed. A remarkable story told by local writer Roderic O'Flaherty in the seventeenth century involved an eel fisher who caught an eel at Cong on Lough Corrib. He secured it through a gill with his belt, which had his knife and purse attached. However, it escaped, carrying away his belt, knife and money. Luckily for him, according to O'Flaherty, the eel was recaptured more than 30 km downstream at Galway (Hardiman 1844; Moriarty 1997).

Many Irish folklore records report explanations for why large eels are found in isolated wells and pools away from rivers. They were thought to have originated from horse hairs that turned into small eels. One informant stated to the folklorist that "when they get too big they leave the hole and make for the river making hoops of themselves—tail in mouth-go like that as long a slime lasts." Some folklore references are to the association of eels with dead animals in water and this is given as a reason for not eating eels. Other reasons why eels were said to be unpopular include suggestions that they had a connection with the devil. It was also suggested that a reason for the aversion to eating eels may be the supposed similarity to snakes, reputed to have been driven from Ireland by St Patrick, or to the many lake monsters mentioned in folk tales (Dunne 2009). In contrast, some entries refer to the fine flavour of eels when grilled, and there are many published accounts of eels being eaten in areas where they were captured in fishing weirs or by other methods.

A folklore report in the records of the Folklore Commission in Dublin mentions practical uses for eel skin in the western Irish Aran Islands, where they were used to make ties for sheep and goats, covers for balls and a gun case. Elsewhere, a folklore reference is made to a giant eel that was killed and from the skin of which dozens of purses were made. Eel skin was also thought to make fine thongs and razor strops (Danagher 1964). According to Evans (1967), eel skins were often used as tough but flexible hinges for wooden flails used in threshing corn.

A wealth of Irish folklore relates to the superstitions of fishing communities. In the case of the Lough Neagh fishing community, a survey by Donnelly (1986) revealed that >40 % of fishers interviewed believed in a variety of superstitions. Many such beliefs were of the type widespread in maritime fishing communities, such as a fear of meeting a red-haired woman on the road when going to fish. There was also a belief that it was unlucky to turn back, for instance if they had forgotten something ashore, and a reluctance to bring the body of a drowned person into a boat (they would prefer to drag it ashore behind the boat). Lough Neagh fishers once used to tie a small piece of flannel to their net, to keep a curse away, and also considered it lucky to place donkey manure in the "tail" of a net. There were many beliefs about fishing in particular parts of the lake at specific times, which may have reflected experiences in previous generations. There was also reluctance to start fishing on a Saturday. However, according to Donnelly (1986), older fishers were typically more superstitious and by the 1970s, many of the traditional superstitions were being increasingly ignored.

Some proverbs mentioning eels were also documented by Irish folklorists including *Bean, muc agus eascú tríur nach féidir a smachtú*, which translated means "A woman, a pig and an eel, three that cannot be controlled"!

Eels and Irish Art

Irish people have been exposed to a variety of artistic images of eels for a long time. Although his views are regarded as being controversial, one author (Kaulin 2006) claimed that grooves on the surface of Irish prehistoric megalithic monuments, such as the Ardristan standing stone and a nearby portal dolmen grave capstone, represent eels. He believes that the eel's heads and other points on the megalithic stones provided the ancient population with information on celestial constellations. Celtic art, seen in many illuminated manuscripts produced in Irish monasteries, often includes strange zoomorphic forms, including some that appear to represent eels. A contemporary print-maker, Keith Kennedy, recently produced a representation of the River Lee monster (*píast*) reputedly driven from the river basin by St Finnbar, and the image is essentially that of a giant eel-like creature (Fig. 2.1).

A painting by Simon Dick (Fig. 2.2a) was used by O'Sullivan and Breen (2007) to illustrate how by visual reconstruction a prehistoric or medieval fishing activity might have looked at a weir on the Shannon estuary. The eighteenth century Irish landscape artist Thomas Roberts (1748–1778) painted a series of landscapes that include rivers and lakes. One, represented in his "A View of Ballyshannon, County Donegal," and two similar paintings, shows the lower section of the River Erne, in which a series of eel fishing weirs can be seen (Fig. 2.2b). That Roberts painting is on public exhibition at the National Gallery in Dublin. A similar drawing, unsigned but attributed to the same artist, is retained in the collections of the National Library, Dublin. The valuable eel fisheries of the lower River Erne were adversely affected by construction of hydroelectric dams in the twentieth century. Some eighteenth and nineteenth century engravings, often used to illustrate travel books, show other Irish fishing weirs. One 1840 example, showing eel weirs in the River Shannon at Athlone, was mentioned by Went (1950, 1951). A painting entitled "Eel weir at Athlone" by the famous Irish landscape artist Paul Henry (1877–1958) depicts a view of the River Shannon, with clear eel-weir stakes, and is on public display in the Athlone Library.

Pauline Bewick (b. 1935) a well-known British-born but Irish resident artist produced an interesting painting called "Woman fish and eel" (Fig. 2.2c). Though dated June 1984–August 2000, this painting was recently sold at an art auction in Dublin and is now in a private collection. Painted in acrylic, water colours and including pen and ink features on paper, it was signed and inscribed by the artist on a label on the reverse side as follows: "Woman Fish and Eel. She contemplates in the green darkness of an overhanging wood sitting in a flowing stream. The eel has always captured my imagination as a most mysterious creature. A local farmer told us an eel will develop in the bog waters far away from any stream from a hair of a horse's tail, I'm sure it's not true but I believe mystifies us human beings in general. I've painted many a picture of women comfortable in nature, it is a wish that we could be so relaxed without fear of being frightened or disapproved of and that we could be at one with nature's creatures as they would be with us. Pauline Bewick."

Sabine Springer, a zoology graduate and artist, provided the Galway-based poet Moya Cannon with a copper etching of eel spears (Fig. 2.2d) as an illustration for a recent poetry collection (Cannon 2011).

A contemporary Northern Irish metal sculptor, Eamon Higgins, recognized the importance of the Lough Neagh eel fishing community. A large roadside steel sculpture of a fisher sitting on the ground with his typical Irish longliner's equipment, baiting a longline, by Higgins is located at Derrytrasna on a main road near the lake's eastern shore (Fig. 2.2e). A modern metal sculpture (Fig. 2.2f), located on the banks of the River Bann opposite the Mesolithic Mount Sandal site, serves to remind people of the importance of lunar periodicity in eel migration. A stone sculpture by Ken Thompson that depicts the warrior monk St Fanahan bearing his battle staff (*Cinn Cathach*), erected in 1989 near Mitchelstown in Co. Cork, has a carved eel on its base.

Fig. 2.2 (**a**) Painting by Simon Dick illustrating how a reconstructed fish trap at Bunratty might have looked. (**b**) Thomas Roberts (1748–1777) painting of the lower River Erne and Ballyshannon with eel fishing weirs in the foreground. (**c**) Woman, fish and eel by Pauline Bewick (from a private collection). (**d**) Copper etching of eel spears by Sabine Springer. (**e**) Metal sculpture by Eamon Higgins of a Lough Neagh fisher baiting a longline. (**f**) Metal sculpture on lower River Bann near Coleraine

The Honan Chapel (Fig. 2.3a) on the campus of University College Cork is regarded as a treasurehouse of the finest arts and crafts available in Ireland during the years preceding its consecration in 1916. Although that time was a turbulent period in Ireland, in the years preceding the declaration of independence from British rule, Celtic Revival influences were developed in the arts and literature. The church's stained glass windows, wood carvings, enamel panels, silver artefacts, printed matter and embroidered cloth items are richly adorned with illustrations of motifs similar to those seen in the illuminated manuscripts and high-crosses of the early Christian period of Irish history. The floor mosaics lead visitors along a remarkable central riverine aisle of the church. The mosaic river, arising from the mouth of a fearsome beast (Fig. 2.3b), has many different types of fish (Fig. 2.3c; Hawkes 2004), all directed downstream towards the sea, which is clearly inhabited by a large but strangely eel-like sea creature (Fig. 2.3d). Also, among the river's fish, an eel is clear (Fig. 2.3c). However, although a scriptural link is made in Latin to the sea creature (*Dracones*) by reference to Psalm 148:7 near the church entrance, the eel is just one element of the biodiverse riverine fish iconography. Eel-like creatures are also in evidence in some early Irish Christian manuscripts and were one of the many zoomorphic motifs used by monastic scribes, silversmiths and sculptors of the "Golden Age" when Irish monasticism provided inspiration and leadership in a wider European context. Irish silversmiths, especially in the early nineteenth century and Celtic Revival period associated with early twentieth century Irish nationalism, often produced work with elongated zoomorphic ornamentation that appears to represent eel-like forms (Bowen and O'Brien 2005). Examples of fish on stone High Crosses are generally of a salmon-like appearance, although putatively eel representations are to be seen on the base of the High Cross of Moone. Although early Irish artists seem to have been familiar with eels, however, little information on eel fishing or associated activities of Irish people in the period can be gained from the study of Irish art.

There are stone sculptures of animal forms on several Irish ecclesiastical and secular buildings. A historical neoclassical Dublin building, the Custom House (Fig. 2.3e), designed by James Gandon (1743–1823) and dating from 1781 to 1791, is adorned by a series of 14 fine allegorical capstones over doors and windows. They represent the Atlantic Ocean and 13 principal rivers of Ireland. The stone heads were made by the sculptor Edward Smyth (1749–1812) and are generally referred to as River Gods of Ireland (Scott 1991). The one representing the River Erne has eels in place of hair and has been represented as line drawings in various old prints. However, it was also used on the Irish £100 note (Fig. 2.3f), which was in circulation from 1928 to 1977, and may be the most widely seen artistic representation of an eel in Ireland.

Eels and Irish literature

Celtic mythology influenced many writers, including those important in Anglo-Irish tradition, such as W. B. Yeats, Lady Augusta Gregory and J. M. Synge, but they rarely made reference to eels in this context or otherwise. Likewise, Lafcadio Hearn (Koizumi Yakumo), whose father was Irish and who spent his boyhood in Ireland and holidayed near Lough Corrib, a lake with abundant eels, does not seem to have made much reference to eels in accounts he published on the folklore of Japan. However, eels were familiar to him, and he noted the "baskets of squirming eels" being carried on the coastal steamer that took him to the Oki Islands on a trip that, like that of his relative Synge to the western Irish Aran Islands, introduced him to the wonderful folktales of his adopted country (Hearn 1894; Ronan 1997).

The ancient stories or sagas of Ireland are divided into four cycles; one of them, the Ulster Cycle, revolves around the heroic warrior *Cúchulainn* who was trained as a warrior by Scathach ("the Shadowy One") on the Scottish Isle of Skye. One well-known tale, the *Táin Bó Cualainge* tells of a cattle raid during the reign of King *Aillíl* and Queen *Meadhbh*, rulers of Connacht, the western Irish province. The queen, though owning a magnificent white-horned bull called *Finbheannach*, desired the famous Black Bull of Cooley (*Donn Cuailigne*) that belonged to *Concobhar mac Neasa*, the King of the northern province of Ulster. The tale tells how she sent her army to do battle with the warriors of Ulster and steal the bull, and describes many encounters with *Cúchulainn*. The tale, set in the first century BC, like other Celtic myths provides extraordinary insight into the folklore of the early Iron Age in Ireland, with lots of details of place names, the Celtic pantheon, and the life style and warfare of the inhabitants of Iron Age Ireland. During the *Táin*, *Cúchulainn* encountered the Goddess of War *Mór-Rioghâin* when he single-handed contended with *Maedhbh*'s army. She initially approached him as a beautiful temptress, but when he rejected her advances and was fighting a great warrior called *Loch*, she took the form of an eel and coiled herself around his feet, causing him to fall. However, he rose and broke the eel's ribs. She later confronted him in other forms, but he managed to rebut her attacks. Finally, she transformed herself into a milking cow and her milk helped him recover from his wounds (Smythe 1988; O hOgain 1991). This is the best-known example of an Irish myth in which eels feature. Some accounts of *Cúchulainn*'s weaponry, which included a harpoon-like javelin that he used to deliver upwards from under water using his foot, to fatally wound an opponent, such as when he fought his old friend *Fer Diad* at a river ford (Carson 2007). In some stories, it is said to resemble a silver eel. Legends such as the *Táin* may have had origins in ancient myths of Mediterranean or Middle Eastern origin, as suggested by the miraculous birth of the two bulls which had fairy cows

Fig. 2.3 Eels and Irish art. (**a**) Honan Chapel, Cork; floor mosaic, with (**b**) river god forming river, (**c**) fish, including eel, moving downstream, and (**d**) river entering sea, with eel-like monsters; (**e**) Dublin Custom House; (**f**) £100 note showing River Erne head, with eels

as mothers (MacKenzie 1907). In Irish mythology they were said to have previously taken the form of ravens, water monsters and humans, then turned into eels. One of the eels went into the River *Cruind*, where it was swallowed by a cow belonging to *Cuailgne*; the other went into the spring of *Uaran Garad* in Connaught, where it passed the belly of a cow of Queen *Maedhbhs* (Squire 1905). Thus, these two important animals in Irish mythology had spent a significant part of their existence in eel form, although this aspect of the *Táin* story is less widely known today. Ancient tales such as these were recorded from the seventh century on by monastic scribes.

According to the *Acallam na Senórach* ("The Colloquay of the Ancients"), a twelfth century narrative recorded in later manuscripts (Ó'Cuív 1983) such as The Book of Lismore (The Book of *MacCárthaigh Ríabhach*), some surviving *Fianna* warriors met St Patrick. In the course of conversations with the saint, one of them, named *Caeilte*, described the food eaten by the *Fianna* as including "Rapid salmons out of *Linnmhuine*, the eels of the noble Shannon; woodcocks of *Fidhrinn*, otters out of the Deel's hidden places" (MacNickle 1933). In addition to the Ulster Cycle, the Fenian Cycle, the Mythological Cycle and the Historical Cycle, Irish sagas have included stories of visions, adventures and travels to imaginary places such as the "Otherworld" and "Land of the Young" (Dillon 1948). One such saga, or "immram," tells of the voyages of the *Uí Corra* for which the text, in the fifteenth century Book of Fermoy, seems to have originated in the middle Irish period (probably the eleventh century). Stokes (1893) edited and translated it. In an extraordinary tale, which involves a pact with the devil and travels in the Otherworld, experiencing visions of hell and wonders, at one stage led by birds that represent the "souls that come on Sunday out of hell," they see three wondrous rivers. These were a river of otters, a river of eels and a river of black swans. The otters, eels and black swans represented devils pursuing the souls of the persons enduring punishment in hell. Knowledge of this saga might in part account for the fear that many Irish people had of eels in the past. Apart from manuscript references, ancient mythological stories such as these also survived in the oral tradition and were told until the early twentieth century by traditional storytellers. An interesting folk-tale, recounted by Joyce (1911), called the "Blind eel Fisher at Clonmacnoise," involved a blind man who could reputedly dive into the River Shannon and return with eels caught between his fingers and toes.

John McGahern, who died in 2006, has been acclaimed widely as one of the most notable Irish writers of the later twentieth century. Many of his stories are set in the Irish midlands (McGahern 1992), where he grew up as the son of a widowed policeman who acquired a farm. One of McGahern's best-known short stories, "Korea," seems to have been set in an area in the upper part of the River Erne. It provides insights into the changing relationship of a father and his son in the complex rural environment of Ireland in the early 1950s, when memories of the bitter Irish civil war still influenced local politics (Cowie 2009). The father and son happen to be eel fishers, and much of the story revolves around their activities on the local lake. McGahern gave few interviews, but in one of them mentioned, when asked about his childhood, that he knew the McMorroughs, one of the traditional eel fishing families in the upper River Shannon (Collinge and Vernadakis 2003). Moreover, he said that he had set eel lines using a boat belonging to the police barracks where his father worked. To him it was a way of "escaping from the house" and unhappy childhood experiences. In fact, eel fishing was, according to the writer, just one of those things that "everybody did." This explains his intimate knowledge of the life of an eel fisher. A feature film "Korea," based on the McGahern short story of the same name, directed by Cathal Black and produced by Darryl Collins in 1995, won the Asta Nielsen Film Award at the 1996 Copenhagen Film Festival. The film includes many references to eel fishing and the loss of a fishing licence by one of the main characters. The fisher and his son are shown setting and lifting longlines on an Irish midlands lake, and the filmmaker also repeatedly shows eels swimming in a typical storage container. Even the simple external latrine at the house of the fisher was full of boxes of earthworms that were used as bait.

Many Irish children have been entertained by the story of Alice in Wonderland, originally written by Lewis Carroll in 1865, which usually has an illustration of old Father William balancing an eel on the tip of his nose (see the UK chapter in this book). Such extraordinary imagination is typical of children's storybooks in the past two centuries. A modern story (Higson 2005) in the Young James Bond Series called "Silverfin," which has impressed Irish boys, tells about a special type of eel that acquired a taste for human blood and that devoured the villain "Meatpacker" at a castle in Scotland! However, not all references to eels in the books read by Irish children are so strange, and they mainly mention as an aside the eels in Ireland's lakes and rivers, or as them being eaten by aquatic birds or otters. Typically, although they may be squeamish about handling eels, Irish children are fascinated by eels and their long-distance trans-Atlantic migrations. In Galway in western Ireland, the children of a primary school recently undertook a project on eels, and in former times, stories of giant eels in lakes served to ensure that children stayed away from dangerous lake shores in the evening; nowadays, however, the eel generally has a more benign image.

Eddie Lenihan a well known West of Ireland storyteller, in a popular story book (Lenihan and Green 2003) tells the story of a hardworking thatcher who witnessed a huge eel rolling past like a wheel, holding its tail in its mouth. The thatcher's wife refused to believe him. Therefore he later catches and cruelly kills the eel to prove to her that he was not confused or drunk and not telling lies. The eel was >3 m long and with "a long mane of hair down its back, just like a horse." Subsequently, the thatcher suffered greatly through injury and could no longer work. He received no sympathy from his neighbours, because they believed the eel was really a person from the Otherworld and was rolling on land as it was passing messages from those that lived beneath one lake to those in another.

Lady Augusta Gregory (1852–1932) was, with her friend the poet Yeats, an important figure in the Irish Literary Revival movement of the early twentieth century. She was a dramatist and folklorist. Her books retell in English stories taken from Irish mythology. She was actively involved in promoting theatrical productions in Ireland and wrote 28 plays (Malone 1924). One of her plays, called "Kincora," was first produced in 1905. It took its name from a historical site near Killaloe on the River Shannon where eels were fished for a long time and which is now an important research location for eels. This tragedy took an eleventh century episode in Irish history for its theme. Among the characters in the play is Gormleith, wife to the High King Brian Boru and formerly wife to Malachi, whom Brian later deposed as High King. A servant, called Brennain, comments that she was seen going upriver in a boat and he says:

> Spearing eels she went, up in the shallows of the river.
> She is not one would take her ease and leave the
> Friday without provision. And there are many
> Not having as much as her, wouldn't walk the
> Road with pride

On her return she gives her eel spear and net to Brennain. Gormleith was from a Viking family that had settled in Dublin. Went (1964) referred to historical accounts that reported the death of Turlough O'Brien, a kinsman of Gormleith's husband, during the battle of Clontarf (1014), when he was trapped by the incoming tide against a fishing weir. King Brian Boru was also slain after the battle by a Viking warrior. According to The Annals of *Tigernach*, there was in 1061 "A hosting by *Aodh Ó Conchobhair* (Aodh O'Conor) into Munster, where he burned Killaloe and demolished the weir of Kincora, and they ate up the salmon that lived in the well of Kincora and afterwards the well was destroyed" (Westropp 1892).

O'Sullivan (2003) wrote a short fictional account of the activities of medieval fishers based at Bunratty in the Shannon estuary, as an introduction to an article in which he reflected on his own personal and professional archaeological experiences there. The strong continuities in work practice over the centuries in which fish traps were in operation, and what he termed "the sense of place," had made a significant impression on the writer. However, he recognized that fictional texts can easily convey false impressions of the manner in which people in the past experienced and perceived the landscape in which they lived.

Eels and the Law in Ireland

The early laws of Ireland, often called the Brehon Laws, are known from a series of law-texts that originated in the seventh–eighth century, which survived (often incompletely and corruptly) in fourteenth–sixteenth century manuscripts (Kelly 1988), and these contribute to our understanding of many aspects of the social history of the island. The Brehon Laws deal with a wide variety of issues and specify penalties. Many of the law-texts illustrate the importance of fish in early Irish economy and, although no law-text specifically dealing with fishing rights still exists, there are many references to issues associated with estuarine or riverine fishing weirs. Early Irish weirs made of stone or wickerwork (stakes and wattle), were called *corae* ("stone wall"), *aire* ("woven fence") or *sód*. These were not all constructed for fish capture and the legal references also relate to structures put in place to divert a current towards a bank to drive a watermill (Kelly 1997). Eel fisheries and watermills have been associated at sites until comparatively recently in Ireland. Fishing weirs were used in Ireland to catch both ascending salmon (*Salmo salar*) and downstream migrating silver eels. Kelly (1997) reviewed the Brehon law-text references to fishing, which mention ancient methods used for fish capture that included spearing, gaffing, strokehauling, tickling, netting, hooking with rod and line and poisoning, as well as the use of fishing weirs. The capture of eels seems to have been done mainly at fishing weirs, although the use of fishing spears or lines may have taken place on a more local and limited scale in lakeshore communities, and the weirs were often also sites at which salmon were fished. Nets used to block the gaps, "eyes," in medieval fishing weirs were similar to those used to the present day and were termed *cochall éisg* in some sixteenth century legal texts. This is a term derived from the Latin word *cuculus* meaning "hood" (as worn by monks), which was anglicized to "coghill" a term still used (Kelly 1997).

There have been various disputes about eel fishing rights in the long and complex history of Ireland. Many were of minor or local significance, and most in the past century involved breaches of bylaws concerning fishing seasons or methods. However, others had greater importance and resulted in consideration of constitutional rights and historical events. Claims to the ownership of the fishing rights to the River Bann and Lough Neagh from the seventeenth century on, resulting from displacement of old Gaelic families, are well documented (Donnelly 1986) and have been discussed recently by Fort (2003). With the development of a lucrative export trade in live Lough Neagh eels to Billingsgate Market in London in the middle of the nineteenth century, the issue of ownership became intense. Local fishers, following cases heard in the British House of Lords and the Dublin High Court, were denied the right to fish freely on the lake, and this encouraged them to resort to illegal fishing. Other legal disputes followed in Northern Irish courts, including an 8-week case in 1962. This led to local conflict with fishery managers and police, which was not resolved until 1971 when a cooperative movement of fishers and local residents succeeded in buying the fishing rights. The Lough Neagh Fishermen's Society Ltd, established under the stewardship of local priest Father Oliver Kennedy, has operated the eel fishery since then (Kennedy 1999). It has been very productive, the largest fishery for wild eels in Europe, and currently is the only river basin in which commercial fishing for eels is now permitted on Ireland. The Society, however, faces great challenges at present. Although the quality of lake eels has been recognized, in the granting of Protected Geographical Indication (PGI) status under EU law, the continued dearth of natural recruitment and the need to ensure the level of spawner biomass escapement required by the River Basin District Eel Management Plan render sustainability of the fishery highly uncertain. The increased reliance on expensive imported glass eels presents serious economic problems (see chapters elsewhere in this book). There are also serious concerns that supplies may not be obtainable from the River Severn or continental Europe within the constraints placed by the EU regulation for restoration of European eel stocks. A further concern expressed by fishery managers in Ireland is that continued importation of live foreign-sourced juvenile eels could result in unforeseen biosecurity problems for the lake and other Irish freshwater habitats.

Another noteworthy legal dispute arose in the case of the River Erne salmon and eel fishing rights when, a few years after the formation of the Irish Free State following signing of a an Anglo-Irish treaty in 1921, the entitlement of the Erne Fishery Company was challenged by local commercial salmon fishers. This led to a District Court, a High Court and ultimately a Supreme Court Case (Moore vs. The Attorney General) in 1927, in which reference was made to Brehon Law, the *Magna Carta*, the nature of Gaelic society prior to the Anglo-Norman conquest of Ireland and historical controversies concerning the Plantation of Ulster. The case encouraged the Irish Free State to legislate important constitutional matters and to establish that the Irish Supreme Court was immune from interference by the Privy Council in London. This legislative change was recognized, but not welcomed, in Great Britain as being constitutional by the Privy Council in 1933. Apart from the implications for other fishery ownership legal cases in Ireland, the decision of the Privy Council was internationally seen as a tremendous advance for Irish sovereignty (Mohr 2002).

The Supreme Court decision in respect of the Moore vs. The Attorney General case had huge implications for management of salmon populations, but commercial fishing for eels was continued on the lower River Erne. Ultimately, the eel fishery was largely destroyed when in the 1950s, the weirs were purchased and removed to allow the Electricity Supply Board to build two hydroelectric dams on the river. On the River Shannon, the Electricity Supply Board acquired statutory responsibility in 1935 for the salmon and eel fisheries, with compensation being paid to fishers, following recognition of the decline in fish stocks that had resulted from the building of a hydroelectric dam at Ardnacrusha in 1929 (Cullen 2002).

The EU Council Regulation (EC Reg. 1100/2007) concerning restoration of European eel stocks required the Republic of Ireland and the Northern Irish Assembly to legislate for eel fishery closures and in respect of other measures contained in the Irish National Eel Management Plan and the International River Basin District Eel Management Plans that involve cross-border river systems such as the River Erne. These have not been welcomed universally, however, and some commercial fishers are currently seeking to challenge the constitutional legality of the bylaws involved.

A challenge by eel fishers in the Shannon region to bylaws banning commercial eel fishing there since 2009 has been rejected by the High Court recently. In a judgement rejecting arguments by James McArdle of the Shannon Eel Fishermen's Association, Mr Justice Daniel Herbert said the purpose of the 2007 regulation was to ensure "effective and equitable" eel-recovery measures. He stated that the two bylaws at issue were not unreasonable or incapable of objective justification by reference to scientific evidence [on causes of depletion of eel stocks], "however lacking in certainty that might be." He rejected arguments that the bylaws meant that the Minister had permanently closed the Shannon river basin area to European eel fishing. Future improvements to eel stocks may allow the reopening of Irish commercial fisheries, of course, but the consensus view among those researching eels is that this will not be at best for many decades because of ongoing poor recruitment in Ireland and the overall poor status of the species throughout Europe.

Eels in Irish Songs and Poetry

Eels are mentioned in popular songs such as the Irish ballad "Nell Flaherty's drake," written in the nineteenth century (O Lochlann 1939) and whose words are said to refer in code to an uprising against the English in Dublin in 1803 by the subsequently executed patriot Robert Emmet. Colourful curses are expressed in the lines

May every old fairy from Cork to Dunleary
Dip him smug and airy in river and lake,
That the eel and the trout, they may dine on the snout
Of the monster that murdered Nell Flaherty's drake.

Moya Cannon, a contemporary Irish poet, included in her collection "Oar" a poem entitled "Holy Well," in which she evocatively captures images from the karstic Burren landscape in western Ireland (Cannon 1990). The final lines of the poem include a brief mention of an eel:

Yet sometimes,
swimming out in waters,
that were blessed in the hill's labyrinthine heart,
the eel flashes past.

Such short, at times almost incidental, references to eels are usually as much as one can expect to find in Irish poetry, with one notable exception. Seamus Heaney, who was born in 1939 in County Derry and who grew up close to Lough Neagh, is recognized as one of the leading poets of his generation. Among the many honours bestowed upon him was the Nobel Prize for Literature in 1995. As a young man he became familiar with the famous Lough Neagh eel fishery, where hundreds of fishers caught yellow eels by longlining and netting. Fishing for silver eels at the Toomebridge weir on the River Bann at the outlet from Lough Neagh was also a familiar sight to Heaney. In many of his poems he mentions eels or eel fishers (Heaney 1998), but perhaps the best known references to eels are in the Heaney poem "A Lough Neagh Sequence," published in 1969, in which he wrote of the famous eel fishing weir:

At Toomebridge where it sluices towards the sea.
They've set new gates and tanks against the flow.
From time to time they break the eels' journey
And lift five hundred stone in one go.

Heaney also gives a wonderful visual sense of fishers gathering earthworms as bait, in his lines

Lamps dawdle in the field at midnight.
Three men follow their nose in the grass,
The lamp's beam their prow and compass.

Likewise, he gives a great sense of the longline setting and lifting activities in lines such as

A line goes out of sight and out of mind
Down to the soft bottom of silt and sand
Past the indifferent skill of the hunting hand.

and

Drawn hand over fist
Where every three yards a hook's missed
Or taken (and the smut thickens, wrist-
Thick, a flail
Lashed into the barrel
With one swing). Each eel
Comes aboard to this welcome:
……….

Although he has lived in Dublin for many years, and has been unwell, he has continued to write about the lake and its eels, for instance in his most recent poetry collection called "Human Chain" (Heaney 2010), which includes the poem "Eelworks" (a term used locally for the eel Fishermen's Cooperative buildings and weir structures at Toomebridge). In this short poem the heavy, old fashioned eel fisher's boats on the lake are described in the lines

Cut of diesel oil in the evening air,
Tractor engines in the clinker-built
Deep-bellied boats,
Landlubbers craft,
Heavy in the water
As a cow down a drain.

Irish Eels as a Natural Resource

Since Mesolithic hunter-gatherers arrived in Ireland around 10,000 years ago, the island's marine and freshwater natural resources have been exploited as a source of food and other material. Relatively little is known about early Irish people, but research has expanded rapidly since the pioneering work in the 1970s at the Mount Sandel site on the River Bann (Woodman 1978; McCartan et al. 2009; Waddell 2010). Analyses of shell middens and vertebrate bones, including fish, birds and mammals, provide useful insights, and it is clear that fishing was important throughout the Irish Mesolithic (10,000–6,000 BP; before present). Faunal remains have only been detected in 13 of 180 Irish Mesolithic sites known to Waddell (2010), so the recent discoveries of organic material such as the fish traps at Clowanstown and in the River Liffey estuary at Dublin are of great interest. Eel bones were recorded in the early Mesolithic sites at Mount Sandel and at Lough Beg just downstream from Lough Neagh. A dugout wooden boat, found on the western shore of Lough Neagh at extreme low water, has been dated to the sixth millennium BC, and these finds, together with records from excavations at Lough Boora in the River Shannon catchment area, show that eels were among the few species of freshwater fish exploited by these earliest inhabitants of the island of Ireland. An early Mesolithic cemetery has been found near Castleconnell, a village on the banks of the lower River Shannon that has been inhabited by eel and salmon fishers for centuries (Carroll and Touhy 1991; O'Sullivan and Breen 2007). However, with the development of farming, it is often assumed that the subsequent Neolithic communities of Ireland "turned their backs on the sea." It is assumed that the Mesolithic people at Mount Sandel chose the site, near a waterfall just up from the estuary, because it was a good place to catch migratory salmon and eels. Likewise, it may have been that progressive use of inland sites rather than coastal ones was in part associated with exploitation of seasonal peaks in availability of these fish species.

Knowledge of inland fishing activities is limited, but many Mesolithic sites (O'Sullivan 1998), such as Lough

Kinale, Lough Derrvarragh, Lough Iron and Lough Allen, are close to areas fished for migrating silver eels until relatively recently. In Clowanstown, Co. Meath, structures interpreted as Mesolithic fishing/mooring platforms were discovered in 2004 and are being investigated carefully. The site, a raised bog area when excavated, would have been a lake in the early post-glacial period. Among the most significant finds were beautifully preserved remains of conical fish baskets (Fitzgerald 2007). Following specialist conservation, some are now on public exhibition in the National Museum in Dublin. The manner in which they were used, or whether some such baskets were used in primitive fishing of weir structures, is not known, but it seems reasonable to speculate that eels could have been among the fish species being captured. Another recent exciting archaeological find was the discovery in Dublin of a wooden fish trap, dated to the late Mesolithic (McQuaide and O'Donnell 2009) and which seems to have been similar to the stake-and-wattle fish traps used to capture salmon, eels and other species from pre-history until modern times. An example of a medieval fishing weir in the River Fergus estuary is shown in the painting reproduced here as Fig. 2.2a.

Hunter-gatherers presumably had good knowledge of seasons and lunar cycles, the determinants of peaks in fish migrations. Prehistoric people in Ireland had an impressive knowledge of astronomical phenomena, and the dates of festivals, determined by lunar and solar cycles, reflected this throughout recorded history. Alignment of many prehistoric monuments such as the Bronze Age passage grave at Newgrange on the banks of the River Boyne to the winter solstice illustrate how accurately a calendar of seasonal fish migrations would have been available in the distant past. There is evidence from analyses of an Iron Age midden near the River Foyle estuary that during a year, seasonally available food was sequentially used, with salmon caught from April to August and eels being in late summer/early autumn (O'Sullivan and Breen 2007). The types of boat used, such as dugout wooden canoes known from the Neolithic to relatively recent times and the possible use of skin-covered, willow-framed boats has been suggested by some. The types of boat used by Irish fishers have varied over the millennia, and even till the late twentieth century, many unusual local designs were in use. In addition to secular community interests in fisheries, monastic communities were, following the introduction of Christianity from the fifth century on, establishing religious and educational centres that were often in estuarine or riverine areas with good fish resources. Apart from fishing weirs, there is direct evidence of the methods used to capture fish, e.g. discoveries of stone weights, fishing hooks and fishing spears. However, from the arrival of the Anglo-Normans in the twelfth century on, when titles to fisheries and disputes were increasingly recorded, the support provided to archaeologists from written and pictorial records increased immensely.

Fish bones recorded during excavations suggest that the medieval inhabitants of the city of Cork included both marine and freshwater species in their diet. Marine gadids such as hake (*Merluccius merluccius*) and cod (*Gadus morhua*) dominate fish bone collections recovered by archaeologists. However, although both European eel and conger eels feature, it seems that the larger marine conger eels were eaten more often than freshwater eels (McCarthy 2003); excavations in Galway, at the Courthouse lane site, revealed a similar pattern. Marine fish such as cod and other gadids dominate both medieval and post-medieval faunal samples, whereas eels and conger eels were of just minor importance. This is an intriguing result, because although there are written records relating to an eel weir in Galway, conger eel remains were more common during the archaeozoological studies (Hamilton-Dyer 2004). Perhaps, therefore, there could be technical problems associated with sampling protocols or relative survival of bones of different fish species in deposits that account for the apparently low level of utilization of eels in Galway then.

Eel Fishing

Young Eels

According to Went (1944), fishing for "eel fry" (glass eels or elvers) for food took place in many Irish rivers in medieval times. Went cites O'Flaherty (1844) as saying that in the River Corrib, at Galway, "an abundance of yelvers or eele frey is taken in Lent time, till they wax black and stiff about May." According to local fishers, glass eels, cooked and pressed to form "elver cakes," were eaten in the twentieth century in some places along the River Shannon. Accounts in old Irish natural history books provide further evidence of juvenile eel consumption, with Boate (1726) mentioning "eel fry" being pressed into cakes and Thompson (1856) describing them as being cooked in milk and pressed into a sort of "cheese." However, it has been illegal to fish commercially for "eel fry" in Ireland since 1842. Glass eels and elvers, as well as small yellow eels (fingerlings), called "bootlace eels" in Ireland, are trapped (Fig. 2.4) as they undertake natural upstream migration and are stocked into lakes. This has been an important fishery management practice in Lough Neagh since 1933, but with declining catches in the past two decades, stocking the fishery now relies increasingly on glass eels imported from the United Kingdom. In two rivers used for hydroelectricity, the Shannon and the Erne, similar conservation trapping has been undertaken since 1959 in an attempt to mitigate for adverse effects on eel population recruitment, and until recently this also involved trapping on adjacent rivers. Special traps are used at the dams, but catches have declined dramatically recently (McCarthy et al. 2008a). Experimental

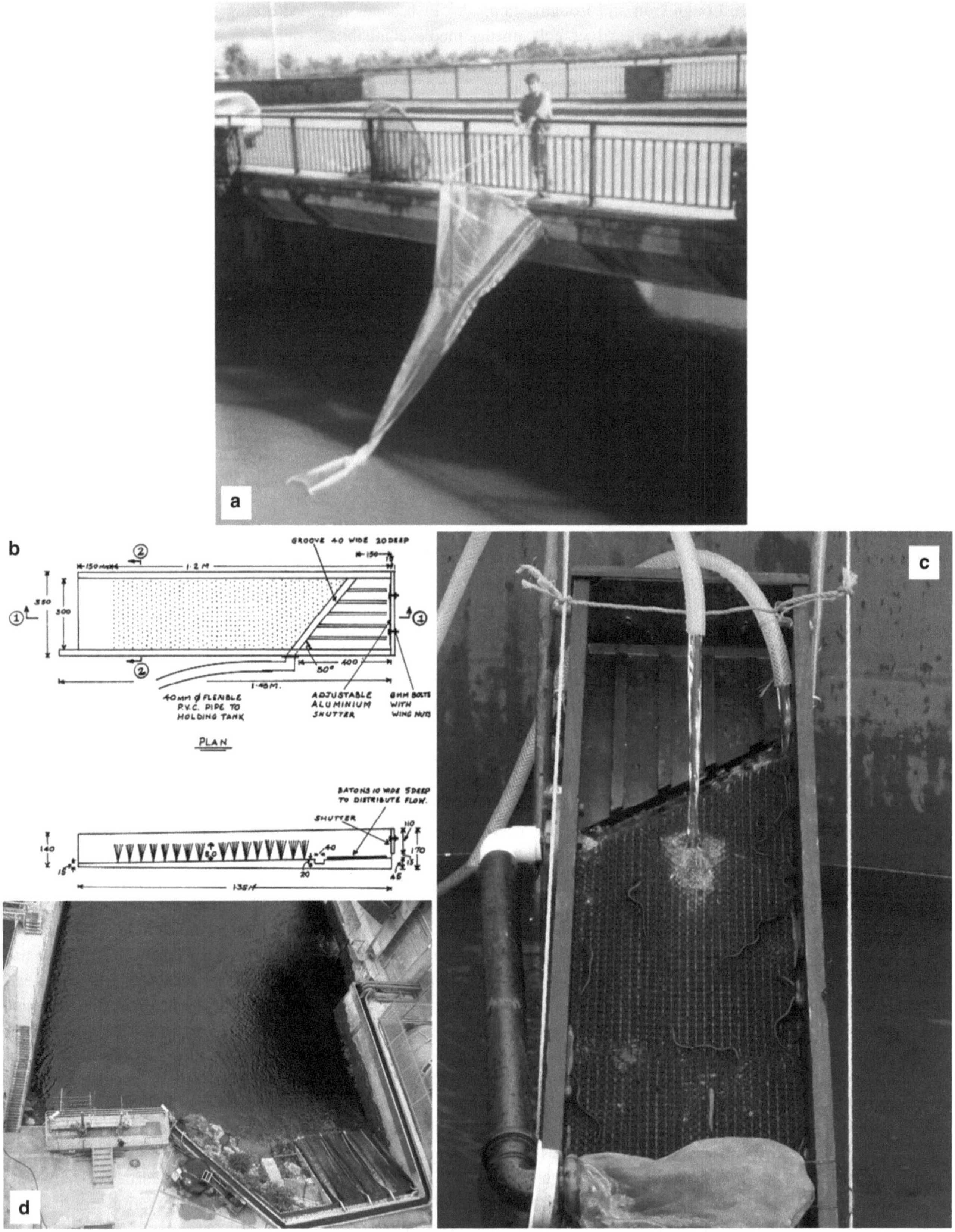

Fig. 2.4 Juvenile eel capture in Ireland. (**a**) Glass eel fishing at Bunratty River estuary bridge; (**b**) the design of O'Leary's (1970) low head elver trap; (**c**) O'Leary trap showing small ("bootlace") eels ascending (**d**) the elver trap at Ardnacrusha hydropower dam, River Shannon

Fig. 2.5 An eel fisher (B. Connell) showing eel spears used by his grandfather and grand uncle on Lough Ree on the River Shannon

use of alternative trap types has been undertaken (McCarthy et al. 1999, 2008b; Matthews et al. 2001) with a view to increasing recruitment to those rivers.

The decreasing recruitment of European eels can be seen in the dramatic decline in the size of the annual catches made at elver traps in the lower River Shannon, where Ardnacrusha catches declined from 4,574 kg in 1980 to 49.4 kg in 2010, and the River Erne, where the corresponding Cathaleens Fall catches declined from 1,352 kg to 93.9 kg.

Yellow Eels

Fishing for eels with spears was legitimate in Ireland until comparatively recently, although the practice had almost died out when Went and Mitchell (1952) published a comprehensive review of Irish fishing spears. However, the technique although then illegal was still in use in some midland lakes in the 1990s and in some estuaries (Fig. 2.5). Eel fishing spears were grouped by Went and Mitchell (1952) into two categories. The first group were transfixing spears called sunspears or rockspears because of the hard substrata upon which they were used. Their handles were often >6 m long and they were used in bright weather when eels could be seen more readily and stalked. Examples of such spears, possibly dating from the early Christian period, have been found in crannogs (lake-dwellings on man-made islands) at several sites in the River Shannon system (O'Sullivan 1998). Sunspears (Fig. 2.6a) were multi-pronged (up to 14) and some had pointed teeth (Fig. 2.6b) between the barbed prongs (Went and Mitchell 1952). A special type of sunspear was used on muddy slobland areas in Wexford Harbour and used by fishers who used ski-like structures (Fig. 2.6c) called "scootches" attached to their boots. Wexford sunspears (Fig. 2.6d) had T-shaped handles <1 m long, or longer if used from a boat, and they were thrust at eels in mud. A pair of outer prongs guided the eel to a single barbed central prong on which the fish was impaled. The other group listed by Went and Mitchell (1952) was termed mudspears, and these were sometimes favoured because they did less damage to the eels. Examples are known from many parts of Ireland and they vary (Fig. 2.6e) in form, some having a circular aperture (Fig. 2.6f) at the bottom of the gap between adjacent tines to facilitate removal of the eel from the spear. Typically there were 3–5 toothed tines. The length of the handle varied and it could be >6 m in the case of those used by bargemen in canals. They were used by blindly thrusting the hand-held spear into lake, river or canal mud and were also sometimes used to gather eels in river meadows to which the fish had moved in large numbers during flood conditions. According to Went and Mitchell (1952), only a single example of an ancient mudspear exists, but future aquatic archaeological research may provide a better knowledge of the antiquity of this fishing method.

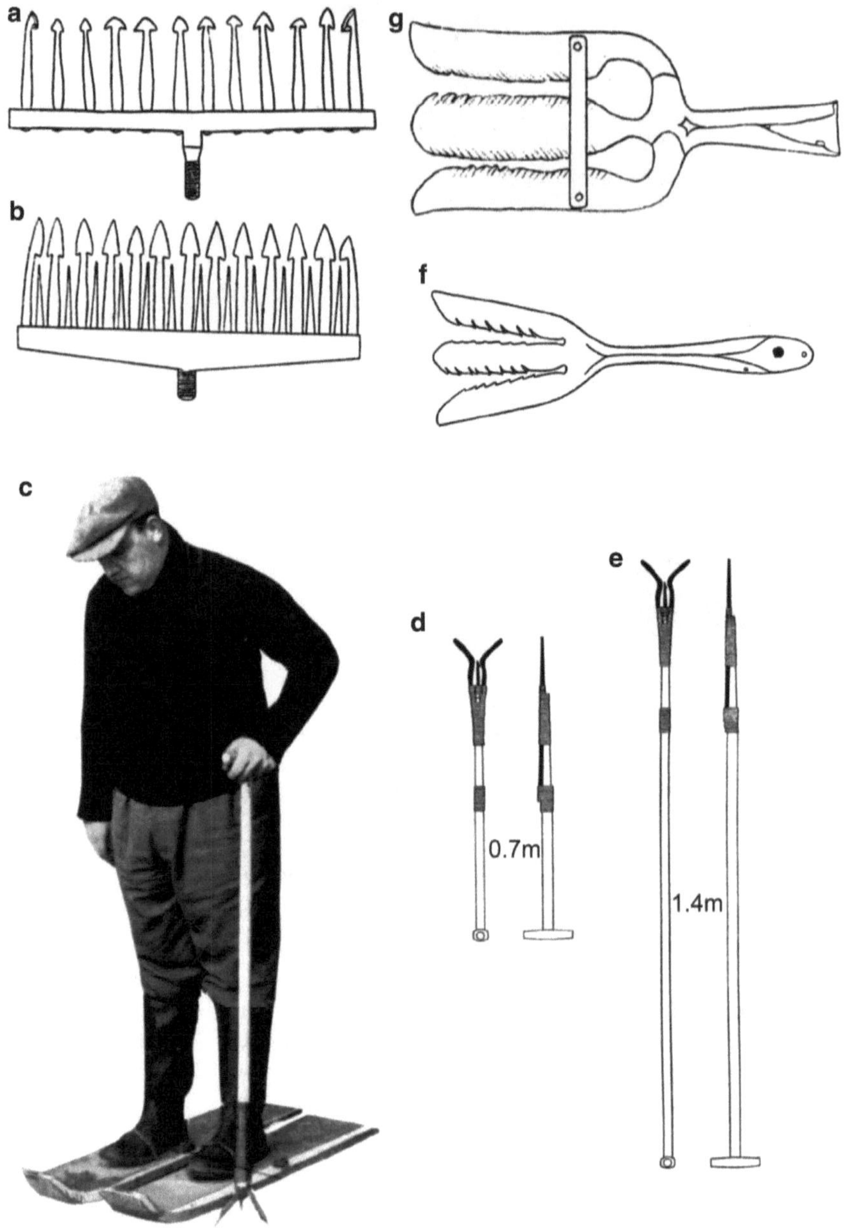

Fig. 2.6 Irish eel fishing spears. (**a**) Sunspear, Lough Ennell; (**b**) sunspear, Lough Key; (**c**) Arthur Went wearing "scootches" and holding a Wexford sunspear; (**d**) short-handled Wexford sunspear; (**e**) long-handled Wexford sunspear for boat use; (**f**) mudspear, Lough Sheelin; (**g**) mudspear, Ballynacarriga (after Went and Mitchell 1952)

Angling for eels has generally not been popular in Ireland, where gamefish such as salmon and trout have generally been the preferred target species and many anglers dislike eels. However, a specialist interest in eel fishing did develop in the twentieth century among relatively few recreational anglers who focused on catching exceptionally large eels. Records compiled annually show that "specimen"-sized fish (i.e. >1.4 kg) were regular catches and that the record size for a rod-caught eel in Ireland was 2.86 kg (6 lb 5 oz). In contrast, the specimen fish weight for conger eels is 18.144 kg (40 lb) and the rod-caught record 32.7 kg (72 lbs). The use of eels as deadbait by anglers trying to catch large pike (*Esox lucius*) seems to have been widespread in the past, with special "flight" arrangements of wire and multiple hooks used to generate a life-like spinning movement of the bait.

Went (1944) described how fishers at a canal in Galway in the first half of the twentieth century used to catch eels using simple equipment by a method called bobbing (or "blobbing"). This involved a large number of earthworms threaded on about 28 m of woollen or hempen thread, which was then rolled into a ball and fastened to about 8–11 m of strong line itself attached to a light pole (Fig. 2.7a). Holding the pole, the fisher lowered the bait into the water and as soon as he felt an eel pulling the worms he rapidly lifted the bait with the eel attached to it onto the bank or into a boat. Went (1944)

Fig. 2.7 Fishing for yellow eels in Ireland. (**a**) Bobbing with earthworms; (**b**) wicker eel pots; (**c**) fykenets; (**d**) electrofishing; (**e**) setting longlines; (**f**) lifting longlines

commented that, though no historical records were available, it appeared to have been more prevalent in earlier years. The practice continued for several decades longer in other parts of the country (T. Callanan, personal communication).

Went (1944) also commented on a peculiar fishing technique seen by naturalist William Thompson in 1834 when he visited Galway. In that case, eels were attracted to garbage and caught on hooks attached to long rods.

The use of summer fykenets (Fig. 2.7c) has been an important standard trapping method for comparative surveys of eel abundance in Ireland, and used to monitor population trends, since the 1960s, when the use of trains of unbaited traps by commercial fishers of yellow eels was recommended by government fishery advisors (Moriarty 1996; McCarthy et al. 1999). These are paired hoopnet pots connected by a leader, linked to form a netted wall of traps. Sets can be deployed randomly for scientific surveys or to target habitats of high density selectively for commercial fishing or targeted sampling. They are now recognized as having scientific limitations associated with the variation in catchability of eels in different habitats and differences in the fishing skills of experimental or commercial crews. Prior to the introduction of fykenets to Ireland, basket traps (Fig. 2.7b) were widely used (Evans 1967), and skilled basketmakers were still producing examples of this type of trap into the early 1980s, as illustrated by Shaw-Smith (1984) and examples on display in the Irish Museum of Country Life. In fishing the mixohaline Lady's Island lagoon, local fishers used to stake fykenets in shallow water perpendicular to the shore. In Waterford harbour, these were replaced by baited pots of the type illustrated by Tesch (2003), deployed in series and held from collapse by hazel rods and with dead-fish baits, in quiet backwater areas of the estuary or attached to whitefish (or sprat) weirs (Went 1959).

Longline fishing (Fig. 2.7e, f) has been a popular eel fishing technique in Ireland since at least the mid-nineteenth century. It may have been used earlier, but good archaeological and historical evidence is not available. It is favoured by the management of the highly successful Lough Neagh fishery, because catch sizes can be predicted accurately and traditional fishers generally regard it as a cost-effective and efficient method. However, many administrative and inspectorate staff in much of Ireland disliked the method intensely because it was used widely in illegal fishing and because of their fears about inclusion of salmonid fish in the longline bycatch. Longline fishing can involve the use of 1,000–2,000 hooks baited with earthworms or fish attached by short lines ("snoods" or "drops") to the main line at regular intervals (e.g. 2 m) and set along the lake bottom using stone weights as anchors. The method has been effective in research surveys of Shannon lakes to examine patterns of eel distribution by depth (Yokouchi et al. 2009) and to estimate eel density/biomass by depletion fishing (McCarthy et al. 1999, 2008b).

In the Lough Neagh fishery, the use of trawls and drift (seine) nets is permitted for the capture of yellow eels. However, the former technique was considered to be disruptive and unsuited to the carefully regulated fishery and is no longer permitted (Rosell et al. 2005).

Electrofishing techniques have been widely used in scientifically surveying streams and small rivers in Ireland (Fig. 2.7d), with assessments of density and biomass obtained by depletion fishing, and in some cases lake shoreline eel population studies have been undertaken using similar protocols (McCarthy et al. 2008b). Electrofishing was also undertaken in lower reaches of the River Shannon, below hydroelectric barriers, for transfer to upper catchment lakes as a stock enhancement measure (McCarthy et al. 1999).

Crushing sacks of the poisonous plant known as Irish spurge (*Euphorbia hiberna*) and placing them in rivers was a traditional method once used for illegal fishing in southwestern Ireland (MacCoitir 2008). The plant's latex has ichthyotoxins that are used to stupefy all fish close by, and although the target may have been salmonids, it is likely that other common species such as eels were also taken and eaten.

Silver Eels

Fishing weirs, for salmon, eels and other species, have been operated for thousands of years. The earliest known example of such a trapping system is a Mesolithic one, recently found in Dublin in the estuary of the River Liffey (McQuaide and O'Donnell 2009). Other Bronze Age and Medieval examples (Fig. 2.8a) have been found in the estuaries of the Rivers Fergus and Shannon in the southwest and in Strangford Lough intertidal areas in northeastern Ireland. At present, however, just three weirs, one on the River Shannon and two on the River Bann, still operate. Many former estuarine fishing weirs, which included ebb and flow types, were used to catch salmon. However, these would also have captured other species of estuarine fish that at some sites would have included eels. Construction methods and materials used differed from place to place. Mitchell (1965) described the remains of an ancient wooden stake and brushwood (wattle) fishing weir at Lough Beg on the River Bann and noted how similar its construction was to the eel fishing weir still operated by the Lough Neagh Fishery at Kilrea (Fig. 2.9) and to the former structure of the well-known Toomebridge eel fishing weir. The modern Toomebridge weir is an impressive mechanized steel structure equipped with special storage containers and hydraulic lifting facilities (Fig. 2.9c). Many eel weirs of the old stake and wattle type (Figs. 2.8a, c) are known to have been fished on Irish rivers, and their history was detailed by distinguished fishery scientist Arthur Went in a series of publications from 1944 to 1969. Among those he described were some that had structures for catching both ascending salmon and descending silver eels at the same location (Fig. 2.8b). Some weir barriers were made of stone, others of wood/brushwood. In the gap, or eye, of a weir, a basket or long coghill net would be set to retain the eels trapped. Many eel weirs caused navigational problems for barges and other boat users of the rivers, however, and there were many disputes about eel fishing for that reason. River drainage, for flood control and other reasons, resulted in the

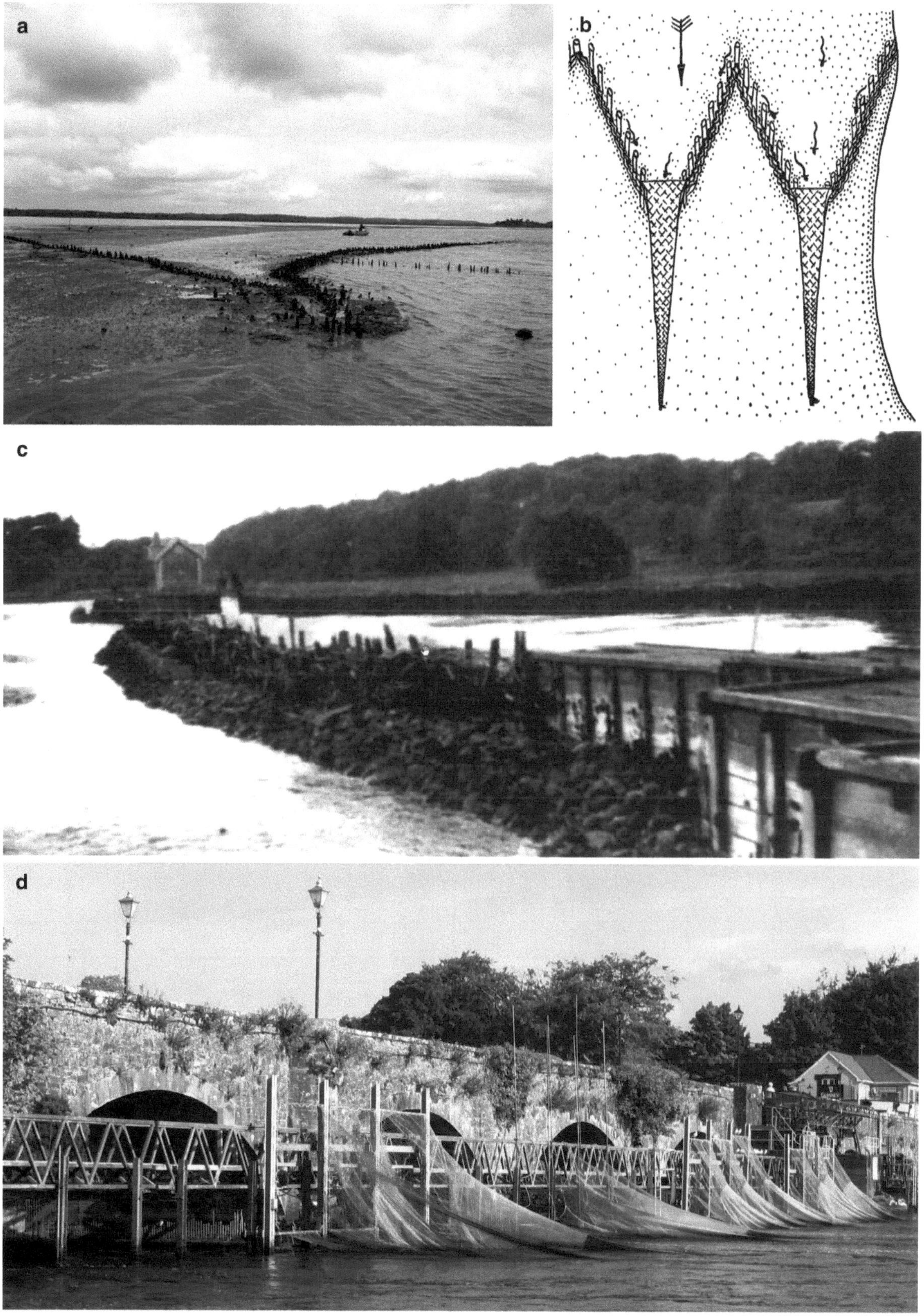

Fig. 2.8 Fishing for silver eels in Ireland. (**a**) Medieval fishing weir, River Fergus estuary; (**b**) reconstruction of a medieval fishing weir at Galway city (after Went 1945); (**c**) former eel fishing weir on the lower River Erne; (**d**) Killaloe eel fishing weir, lower River Shannon

Fig. 2.9 The Lough Neagh Fishery. (**a**) Preparing a line for eel fishing, c. 1915; (**b**) Northern Ireland Agriculture Minister at Lough Neagh Eel Fishermen's Cooperative with Father Oliver Kennedy (photo courtesy Tyrone Times); (**c**) Toomebridge weir; (**d**) Lough Neagh Fishermen's Cooperative buildings; (**e**) an elver trap at the Cutts weir on the lower River Bann; (**f**) gate with Lough Neagh Eel Fishermen's Cooperative logo

removal of many weirs (Delany 2008). Also, the wooden structures required much annual maintenance and if they ceased to be used, they disappeared rapidly, leaving no indication of their location. Old property title deeds, documents concerning fishing weir leases, estate records, wills and old ordnance survey maps are, however, important historical sources of information on weirs. References to old weirs in local history books, old travel books and some public

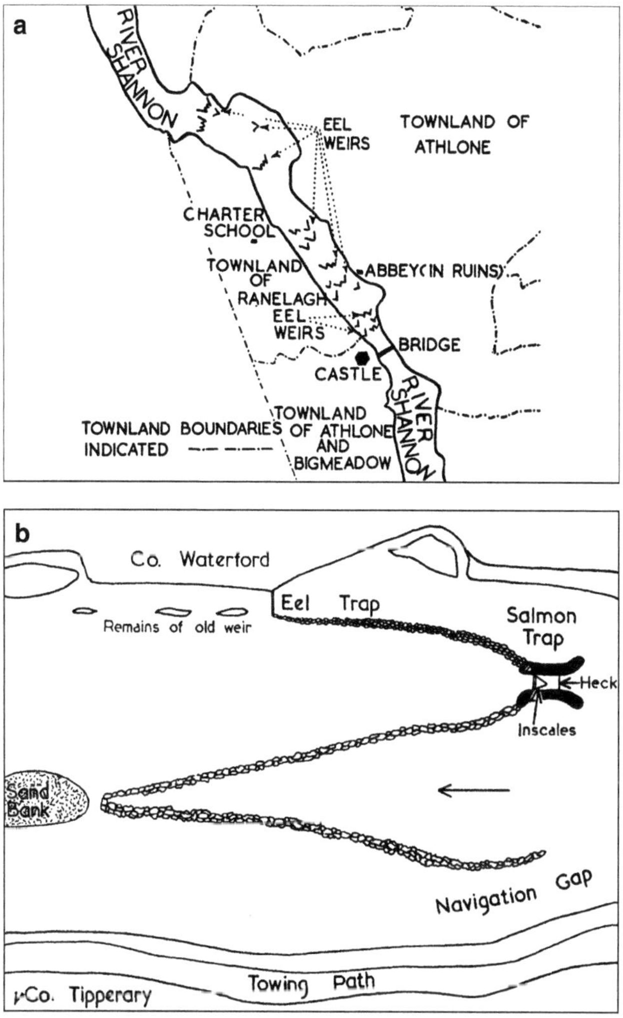

Fig. 2.10 Old Irish fishing weirs. (**a**) Location of numerous old fishing weirs at Athlone on the River Shannon (after Went 1945); (**b**) fishing weir on the River Suir, used for capture of both salmon and eels (after Went and Mitchell 1956)

records also provide a sense of the scale and value of such eel fishing weirs (Fig. 2.10). Modern aquatic archaeological research (O'Sullivan 1998; O'Sullivan and Breen 2007) also contributes to knowledge of former fishing weirs.

Hardiman (1844) and Went (1944) described the Galway fishery and the history of eel fishing weirs (Fig. 2.8b) along the lower section of the River Corrib, which was first mentioned in official documents in 1283. At that time the salmon fishery was valued at £11 and the eel fishery at £10. The fishery was bought in the late twentieth century by the Irish government as a measure to conserve local salmon stocks, however, and at that time it was said that the eel and salmon fisheries were of approximately similar value in terms of their commercial catches. The value of eel fisheries controlled by religious orders is well documented, especially when during the Reformation the dissolution of the abbeys resulted in the transfer of lands and fishing rights to those favoured by the British monarch and his Irish representatives. At Galway, eel fishing weirs were operated by monks from two abbeys in the city, as well as by monks from as far away (29 km) as Abbeyknockmoy. Some leases of eel fishing weirs were paid by presentation of a portion of the catch to the owner, others by cash payment for long-term lease. Operators of fishing weirs were only entitled to block a portion of the river (e.g. one-sixth).

Several fishing weirs were operated at Castleconnell, on the lower River Shannon, until the 1990s to catch silver eels. These were the last of the privately operated weirs that were common along the Shannon in the nineteenth century. The development of the hydroelectricity generating station at Ardnacrusha resulted in a decline in salmon and eel stocks, and the company acquired the fishing rights and compensated fishers for their loss. However, in addition to its own eel weirs at Killaloe and Athlone, a few independently operated fishing weirs were authorized to fish for silver eels, and the Castleconnell ones were among these. A series of more than ten weirs had been fished on the Shannon downstream of Killaloe in the nineteenth century. A colourful account of the Castleconnell area and its eel fishers was published by Carroll and Touhy (1991). In the first half of the twentieth century, up to 30 men were employed in autumn/winter in fishing or fishery-related activities, and catches were good. The best night's fishing at that time involved a catch of 12 t. The eels were dispatched by rail and boat to the Billingsgate Market in London.

The eel weir operated by the Electricity Supply Board at Killaloe on the lower River Shannon (Fig. 2.8d), initially commercially but now for research and conservation, is a steel-framed structure attached to a historical bridge. Silver eel migrations have been monitored there for decades, providing information on the environmental factors affecting seasonality and population dynamics of silver eels (McCarthy et al. 1999, 2008b; Cullen and McCarthy 2003). The site is also an important location for ongoing research focused on eel management plan data requirements. It was also used from 2000 to develop a silver eel trap and transport programme in respect of the Ardnacrusha dam downstream (McCarthy et al. 2008a). Catches, made with winged river nets, at upper catchment conservation fishing sites are also being monitored.

Many weirs were primarily intended for guiding water to watermills but secondarily used for fish-trapping. In the early twentieth century, special eel traps of the type illustrated by Tesch (2003) were common at mills in some parts of Ireland, and one was operated in Lough Rea, Co. Galway, until recently. These effectively screened all water entering the mill and diverted eels to a storage container.

Silver eels have also been captured in Ireland using winged river nets and large D-framed fykenets set at lake outlets. At the outlet of lower Lough Erne (Fig. 2.11a, b), a complex silver eel fishing system is still operated as a silver

Fig. 2.11 (**a**) Lower Lough Erne with outlet to River Erne in the distance. (**b**) Map of the lower Lough Erne outlet silver eel fishery. (**c**) Researchers examining catches of silver eels for tagging with Passive Integrated Transponder (PIT) tags. (**d**) Fishing crew transferring silver eels to storage box

eel conservation trapping programme. It involves extensive netting held vertically by steel poles in which there are net traps with funnels to prevent eels escaping, and these are operated in conjunction with a series of winged river nets set in series along a deeper navigation channel. That site and an experimental one at Roscor Bridge ~0.5 km downstream are key research sites for monitoring populations of silver eels migrating downstream towards the river section used for generating hydroelectricity (Fig. 2.11c, d). Other sites upstream are fished with winged river nets too, as part of the conservation strategy.

All Irish eel fisheries, except for Lough Neagh/River Bann, were closed in 2009 as part of the Irish and Northern Irish eel management plans, and a system of trapping and transportation (Fig. 2.12a) of silver eels was a required action in respect of the three main rivers (Rivers Erne, Shannon and Lee) used for hydroelectric generation. A series of sites is now fished on all three rivers and their catches, closely monitored, are transported to release points located close to the estuaries. Since 2000, 87.65 t of silver eels trapped at Killaloe on the River Shannon have been transported safely downstream, including most of the 77.18 t transported down the catchment in the past 3 years. The eel management plan specified that 30 % of the annual silver eel run in the River Shannon should be transported and this was achieved in all the past 3 years (35.2 %, 37.9 % and 39.2 %). The silver eel trap and transport programme on the Erne has thus far not managed to capture the quantities specified in the management plan, but this seems to be explained by population studies that suggest that the River Erne eel population is smaller than assumed when the relevant management plan was being compiled. In the initial three silver eel migratory seasons (2009/2010–2011/2012), a cumulative total of 53.95 t of eels was transported downstream and released below the two hydropower dams on the lower section of the River Erne. In the case of the River Lee, with its large man-made reservoirs,

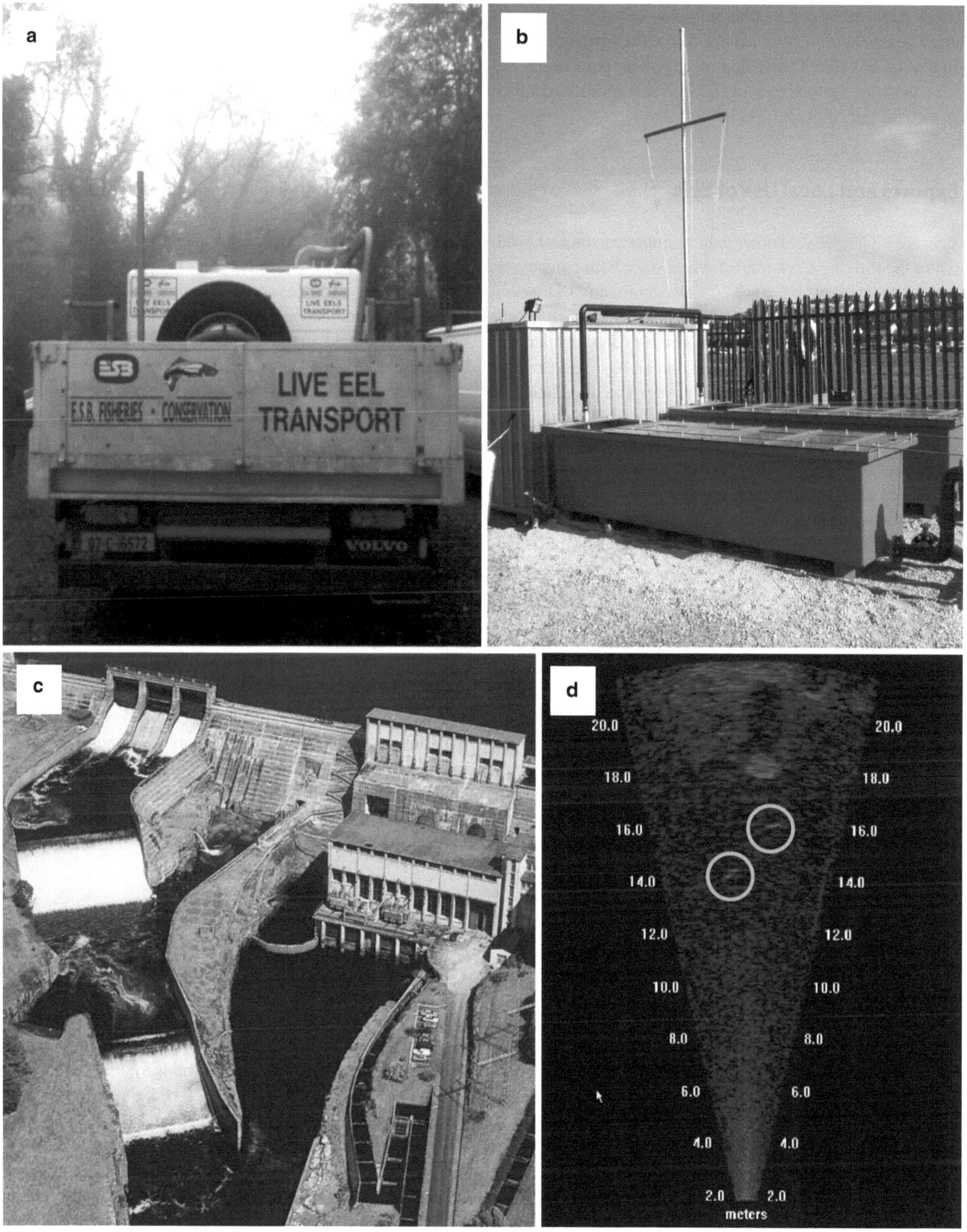

Fig. 2.12 Eel conservation and research in Ireland. (**a**) Vehicle used for transporting silver eels downstream of hydropower stations; (**b**) Migromat technology® being tested at Killaloe; (**c**) spillage at Cathaleen's Fall hydropower station on the River Erne; (**d**) DIDSON® ultrasound camera images of silver eels migrating downstream at Roscor Point, River Erne

it has been necessary to trap pre-silver eels using fykenets and release them to river habitat below the lowermost dam. In that case, where a specified management plan target was 0.5 t annually, the total quantity trapped and transported varied from 0.08 t in 2009 to 0.73 t in 2011.

Exports and Local Use of Eels

Eels are no longer sold other than as imported smoked product in the thriving Cork market where once both conger eels and freshwater eels were sold (O'Drisceoil and O'Drisceoil 2011). A much wider variety of fish is eaten by Irish people nowadays than in the past, when for many centuries fish was regarded as typical of a penitential diet on days specified for special religious observance and fasting. The association of many eel fisheries with religious abbeys, especially in medieval and post-medieval times, may reflect external influences and cultivated tastes. Eels are easily stored alive and could have been an important food for that reason and because of the restrictions on eating meat on days on which religious fasting was prescribed. Eels captured in Ireland in the nineteenth and twentieth centuries were mostly exported alive by train and boat to the Billingsgate Market in London. Recently, however, they were mostly exported alive to The Netherlands. Eels were cooked in various ways in Ireland and also smoked for local consumption and export. In the past, eels were eaten in large quantities in areas with extensive silver eel fishing weirs; for example, in the lower River Shannon, eels were once derisively called "Killaloe Bacon" (i.e. everyday food in that town). Eel skin was valued in Ireland in former times, being used as a flexible joint in flails used for threshing corn as well as for making small leather goods. The closure of Irish eel fisheries, with the exception of the River Bann and Lough Neagh, has led to a decline in interest in eels as a natural resource, however, and emphasis is now placed on conservation.

Research on Eels in Ireland

The ecology of eels in Ireland has been studied in Ireland for more than 50 years, with studies on growth rate, feeding habits, predators, parasites and migration patterns providing a general framework for eel management (e.g. Parsons et al. 1977; Moriarty 1978, 2003; Poole and Reynolds 1996; Doherty and McCarthy 1997; McCarthy et al. 1999, 2009; Matthews et al. 2001; Cullen and McCarthy 2000, 2003, 2007; Rosell et al. 2005). However, the need to provide scientific advice in respect of the eel management plans produced in response to the EU Regulation on restoration of eel stocks has increased the need for research. Eels are widespread in Ireland, in coastal, estuarine, lagoon, riverine and lacustrine habitats. Accordingly, research tasks and methodologies have been varied.

To date, research on coastal eel populations has been of a limited or site-specific nature (Harrod et al. 2005; Arai et al. 2006; Morrissey and McCarthy 2007), but major surveys are currently being undertaken by Inland Fisheries Ireland in the Waterford and Wexford estuaries. Information on eel biology in Ireland is now considered to be of high priority. The determination of population parameters for open estuarine habitats, where in some instances apparently good numbers of eels still exist, is proving to be difficult, however. Consequently, and because of potential movements of eels between riverine and estuarine areas, estimating the contribution such stocks make to spawner biomass escapement is not possible.

The determination of natural production and spawner escapement biomass from selected Irish rivers is also a major part of current eel population research in Ireland. This is information needed for evaluating eel management plan actions such as fishery closure and mitigation measures for hydropower. Estimation of spawner escapement biomass has been possible (R. Poole, unpublished) on the small Burrishoole River by efficient trapping, but has been more difficult on larger rivers such as the Rivers Bann (Rosell et al. 2005), Erne and Shannon. However, there has been considerable success in the case of the latter two rivers in the past 2 years by the use (Fig. 2.12d) of modern technologies (DIDSON® ultrasound camera counts, hydroacoustic surveys, acoustic telemetry) combined with investigations at experimental fishing weirs (population structure analyses, mark-recapture estimates of population size, etc.). Likewise, studies on parasite burdens of seaward-migrating silver eels, on their fat levels and maturation status, is providing better knowledge of the quality of potential spawners leaving Irish waters. This is a matter of increasing concern given the spread of the swimbladder parasite *A. crassus*.

Research is also in progress to facilitate the development of better measures for mitigating the effects of hydropower generation on migrating eels (Fig. 2.12c). The aim is to determine whether the trap and transport measures currently in place can be replaced or supplemented by new-generation protocols, and the technologies being evaluated include (Fig. 2.12b) the Migromat early warning system® and deflection screens. Also, route selection is being monitored on the Rivers Erne and Shannon by telemetry.

A major objective of current Irish eel research is the development of effective modelling tools for managing Irish eel populations in support of the international efforts being made to recover the European eel stock. Irish scientists participate in European research programmes such as the Eeliad project, and eels from Irish rivers have been used in satellite-tracking and other telemetry studies of silver eels during their spawning migration (Aarestrup et al. 2009).

Overview

A wide ranging review such as attempted here, even when focused on a topic with defined parameters—two species (European eels and humans), a clear geographic area (Ireland), time-frame (since Pleistocene glaciation, >10,000 years) and scholarly disciplines (archaeology, history, art/literature, folklore, natural resource management, ecology)—has proven more challenging and complex than anticipated. Looking forward in time, the thin threads of knowledge that allow for understanding the ecological, social and cultural links between humans and eels provide limited perspectives. However, despite the biocomplexity and methodological limitations, a sense of the uniqueness of the eel/human association can be appreciated by reference to the diversity of avenues explored. Moreover, the need for a more systematic, comprehensive review and further research is apparent. Urgency seems appropriate in this regard given the decline in the global stock of European eels, the closure of Irish and many other eel fisheries, the continuing threats posed by environmental change, including invasive pathogens, and societal changes that result in the loss of place- and folklore in modern Ireland. Exchange of information on eels and humans may serve to highlight the global problem facing anguillid eels, and also highlight the need for deeper understanding of the interactions between humans and these fascinating fish. Humans, once simply a predator on eels, now influence them adversely in a variety of ways.

Eel conservation objectives will be better served in Ireland by a deeper understanding of the historical and cultural importance of the species, especially if this serves to draw public attention to the environmental issues involved. Ireland's poets, writers, artists and film-makers can contribute by enlightening Irish people lacking direct connection to eels or their aquatic habitats. However, in this age of digital media, one in which educational courses are increasingly narrow or vocationally defined, there is also a need to remind educationalists how fundamentally important it is to have a multidisciplinary approach to many of life's problems. The old term "natural history", now largely redundant, encompassed a diversity of what is now regarded as specialist disciplines. Perhaps, the future of eels depends on us returning to our intellectual roots and not being afraid to cast our net widely in the search for a deeper knowledge of this topic. Analysing the history of the eels and humans story in an Irish context has revealed many unresolved issues, so, in attempts to look forward, it is clear that the future of the eels and of human interactions with them is characterized by great uncertainty.

The nocturnal activity of eels may have contributed greatly to the mystery associated with them in Ireland. Their association with lake monsters in mythology and folklore, as well as their snake-like appearance, seems to have created a negative popular image. Landscape artists, writers and the general public rarely see eel fishers at work because they work largely during the hours of darkness. However, because of the dramatic collapse in eel populations and because of the publicity associated with the implementation of the national eel management plan, which was required by EU legislation, a new awareness of the uniqueness and importance of eels is being created. The future of eels in Ireland will depend in part on the measures being implemented to enhance stocks, to reduce anthropogenic mortality and to increase spawner escapement to sea. This is being done in a cooperative international context and, like the implementation of the EU Water Framework Directive, on a river basin district basis (Fig. 2.13). However, because of the panmictic nature of the eel, the ultimate success or failure of the EU plan for the recovery of stocks of the European eel will depend on international cooperation across the full geographic range of the species.

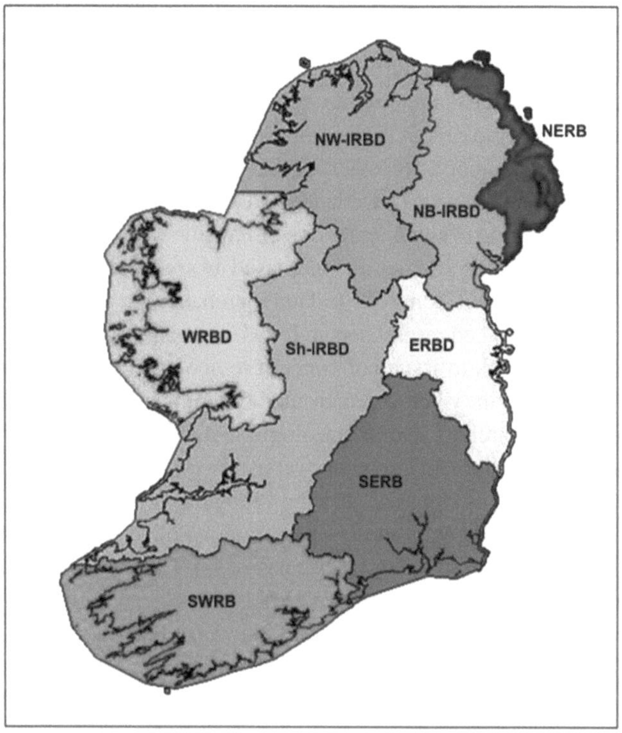

Fig. 2.13 Map of Ireland showing the River Basin districts used for the EU Eel Management Plan and implementation of the EU Water Framework Directive

The specialized discipline of modern archaeology contrasts greatly with the activities of nineteenth century antiquarians who regularly sought to interpret ancient structures by reference to mythology or early manuscript texts on history. Waddell (2005) described the development of Irish archaeology and illustrated how archaeological interpretations were influenced strongly in this way into the early twentieth century. Therefore, while the myths, early historical documents and folklore can give some fascinating insights into the perceptions and activities of early inhabitants of Ireland with respect to eels, it is also important to recognize the limitations that apply. Recent advances in the archaeology of Ireland's inland waters, wetlands and estuaries are improving understanding of the interactions of humans with eel populations. New technologies and increased awareness of the need to research and conserve Ireland's archaeological heritage have encouraged researchers to focus on questions that relate in many ways to issues of concern to ecologists. There are many ways in which research methodologies are paralleled by eel researchers and archaeologists, such as in the use of acoustic equipment for underwater surveys, the generation of GIS-based databases, and the recognition of the importance of dealing with these topics in terms of river corridors and river basin districts (McNeary 2011). Likewise, just as nodal points (e.g. fords or lake outlets) have been productive research areas for archaeologists, they have been of particular importance in eel population studies. Hence, many of the river or lakeshore Irish Mesolithic sites that have provided the greatest insights into the early inhabitants of the island are well known to Irish eel researchers and to Irish eel fishers. The continuity of practice, highlighted by O'Sullivan (2007), in which wetland and aquatic areas have been exploited may well have extended way longer than is generally appreciated, and the near terminal decline in the eel fishing community may well result in the loss of a very important part of Irish folk memory.

Acknowledgements The advice and assistance provided by the following in compiling this chapter is gratefully acknowledged: Criostóir MacCárthaigh, Con Manning, Michael O'Connell, Shane Mawe, Brendan Delany, Dariusz Nowak, Ruairí MacNamara, Fintan Egan, Michael Guiry, Aidan O'Sullivan and members of staff at the National Gallery of Ireland and National Museum of Ireland, and artists Simon Dick, Pauline Bewick and Sabine Springer, and especially Anne Bateman.

References

Aarestrup K, Økland F, Hansen MM, Righton D, Gargan P, Castonguay M, Bernatchez L et al (2009) Oceanic spawning migration of the European eel (*Anguilla anguilla*). Science 325:1660

Arai T, Kotake A, McCarthy TK (2006) Habitat use of European eel *Anguilla anguilla* in Irish waters. Estuar Coast Shelf Sci 67:569–578

Boate G (1726) A natural history of Ireland. G. and A. Ewing, Dublin

Bowen JR, O'Brien C (2005) Cork silver and gold. Four centuries of craftmanship. Collins, Cork

Cannon M (1990) Oar. Galley, Loughcrew, Oldcastle

Cannon M (2011) Hands. Carcanet, Manchester

Carroll J, Touhy P (1991) Village by the Shannon. The story of Castleconnell and its Hinterland. Private publication, Limerick

Carson C (2007) The táin. A new translation of the táin bó Cuailgne. Penguin Books, London

Collinge L, Vernadakis E (2003) John McGahern – b. 1934. J Short Story Eng 41:15, online 2008

Cowie D (2009) "Korea" by John McGahern. J Short Story Eng 53:2–9 (online 2010)

Cullen R (2002) Some notable design features of the design and operational history of Ardnacrusha. In: Bielenberg A (ed) The Shannon scheme and the electrification of the Irish free state. Lilliput, Dublin, pp 138–154

Cullen P, McCarthy TK (2000) The effects of artificial light on the distribution of catches of silver eel, *Anguilla anguilla* (L.) across the Killaloe eel weir in the lower River Shannon. Biol Environ Proc R Ir Acad 100B:165–169

Cullen P, McCarthy TK (2003) Hydrometric and meteorological factors affecting the seaward migration of silver eel (*Anguilla anguilla*) in the lower River Shannon. Environ Biol Fishes 67:349–357

Cullen P, McCarthy TK (2007) Eels (*Anguilla anguilla* (L.)) of the lower River Shannon, with particular reference to seasonality in their activity and feeding ecology. Biol Environ Proc R Ir Acad 107(2):87–94

Danagher K (1964) Irish customs and beliefs. Mercier, Cork

Delany R (2008) The Shannon navigation. Lilliput, Dublin

Dillon M (1948) Early Irish literature (Celtic studies). Four Courts, Dublin

Doherty D, McCarthy TK (1997) The population dynamics, foraging activities and diet of great cormorants (*Phalacrocorax carbo* L.) in the vicinity of an Irish hydroelectricity generating station. Supplemento alle Ricerche di Biologia Selvaggina 24:133–143

Donnelly DJ (1986) On Lough Neagh's shores. A study of the Lough Neagh fishing community. The Donnelly Family, Galbally, Co, Tyrone

Dunne J (2009) Irish lake marvels: mysteries, legends and lore. Liberties, Dublin

Edwards R, Brooks A (2006) The island of Ireland: drowning the myth of an Irish land-bridge. In: Davenport JL, Sleeman DP, Woodman PC (eds) Mind the gap. Post-glacial colonization of Ireland, pp 19–34. Special Supplement to The Irish Naturalist's Journal

Evans EE (1967) Irish heritage. The landscape, the people and their work. Dundalgan, Dundalk

Fitzgerald M (2007) Catch of the day at Clowanstown, Co. Meath. Archaeol Ireland 21(4):12–15

Flanagan D, Flanagan L (2004) Irish place names. Gill and Macmillan, Dublin

Fort T (2003) The book of eels. Harper Collins, London

Hamilton-Dyer S (2004) Bird and fish bone. In: Fitzpatrick E, O'Brien M, Walsh P (eds) Archaeological excavations in Galway city. Wordwell, Bray, pp 609–626

Hardiman J (ed) (1844) O'Flahertys "A chorographical description of West ir Iar-Connaught". Irish Archaeological Society, Dublin

Harrod C, Grey J, McCarthy TK, Morrissey M (2005) Stable isotopes provide new insights into ecological plasticity in a mixohaline population of European eel. Oecologia 144:673–683

Hawkes J (2004) The Honan chapel. An iconographic excursus. In: Teehan V, Hecket EW (eds) The Honan chapel. A golden vision. Cork University Press, Cork, pp 105–131

Healy B (2003) Coastal lagoons. In: Otte M (ed) Wetlands of Ireland. University College Dublin Press, Dublin, pp 57–78

Heaney S (1998) Opened ground. Poems 1966–1996. Faber and Faber, London

Heaney S (2010) Human chain. Faber and Faber, London

Hearn L (1894) Glimpses of unfamiliar Japan. Tuttle Publishing, Tokyo

Higson C (2005) Silverfin. Puffin Books, London

Joyce PW (1911) The wonders of Ireland and other papers on Irish subjects. Longmans Green, London

Kaulin A (2006) Stars, stones and scholars. The decipherment of Megaliths as ancient survey of the earth by astronomy. Trafford Publishing, Indianapolis

Kelly F (1988) A guide to early Irish law. Dublin Institute for Advanced Studies, Dublin
Kelly F (1997) Early Irish farming. Dublin Institute for Advanced Studies, Dublin
Kennedy O (1999) The commercial eel fishery on Lough Neagh. Fish Bull (Dublin) 17:27–32
Lenihan E, Green CE (2003) Meeting the other crowd. The fairy stories of hidden Ireland. Gill and McMillan, Dublin
Logan P (1980) The holy wells of Ireland. Colin Smythe Ltd, Buckinghamshire
MacCoitir N (2008) Irish wild plants. Myths, legends and folklore. Collins, Cork
MacKenzie D (1907) Egyptian myth and legend. Gresham Publishing, London
MacNeill M (1962) The festival of Lughnasa. A study of the survival of the Celtic festival of the beginning of harvest. Oxford University Press, Oxford
MacNickle MD (1933) Beasts and birds in the lives of the early Irish saints. Ph.D dissertation, University of Philadelphia
Malone J (1924) The plays of lady gregory. Studies Ir Q Rev 13:247–258
Matthews M, Evans D, Rosell R, Moriarty C, Marsh I (2001) Erne eel enhancement programme. Final report, Northern Regional Fishery Board, Ballyshannon
McCartan S, Schulting R, Warren G, Woodman P (eds) (2009) Mesolithic horizons, vols 1 and 2. Oxbow Books, Oxford
McCarthy M (2003) Faunal remains. In: Cleary RM, Hurley MF (eds) Cork city excavations 1984–2000. Cork City Council, Cork, pp 375–389
McCarthy TK, Cullen P, O'Connor W (1999) The biology and management of River Shannon eel populations. Fish Bull (Dublin) 17:9–20
McCarthy TK, Blaszkowski M, Grennan J, Egan F, Cullen P, Nowak D (2008a) Population biology of European eel (*Anguilla anguilla*) in Lough Derg, an Irish mesotrophic lake ecosystem. In: Phillips B, Megrey BA, Zhou Y (eds) Proceedings of the 5th world fisheries congress, Yokohama, American Fisheries Society, Bethesda, pp 279–280
McCarthy TK, Frankiewicz P, Cullen P, Blaszkowski M, O'Connor W, Doherty D (2008b) Long-term effects of hydropower installations and associated river regulation on River Shannon eel populations: mitigation and management. Hydrobiologia 609:109–124
McCarthy TK, Creed K, Naughton O, Cullen P, Copley L (2009) The metazoan parasites of eels in Ireland: zoogeographical, ecological and fishery management perspectives. Am Fish Soc Symp 58:175–187
McGahern J (1992) The collected stories. Faber and Faber, London
McNeary R (2011) Riverine archaeology in Northern Ireland: an evaluation. Nautical Archaeol 40:162–170
McQuaide M, O'Donnell L (2009) The excavation of Late Mesolithic fish trap remains from the Liffey estuary, Dublin. In: McCartan S, Schulting R, Warren G, Woodman P (eds) Mesolithic horizons, vols 1 and 2. Oxbow Books, Oxford, pp 889–894
Mitchell NC (1965) The lower bann fisheries. Ulster Folklife 11:1–32
Mohr T (2002) Law without loyalty – the abolition of the Irish appeal to the privy council. Ir Jurist 37(1):187–226
Moriarty C (1978) Eels. A natural and unnatural history. David and Charles, London
Moriarty C (1996) Variation in numbers of the eel *Anguilla anguilla* captured by constant effort in an Irish lake, 1881–1994. Ecol Freshw Fish 14:148–152
Moriarty C (1997) Fish and fisheries. In: Foster JW, Chesney HCG (eds) Nature in Ireland. A scientific and cultural history. Lilliput, Dublin, pp 283–298
Moriarty C (2003) A review of eel fisheries in Ireland and strategies for future development. Am Fish Soc Symp 33:217–224
Morrissey M, McCarthy TK (2007) The occurrence of *Anguillicola crassus* (Kuwahar, Nimi and Hagaki, 1974), an introduced nematode, in an unexploited western Irish eel population. Biol Environ Proc R Ir Acad 107B:13–18
O hOgain D (1991) Myth, legend and romance. An encyclopaedia of the Irish folk tradition. Prentice-Hall, New York
O Lochlann C (1939) Irish street ballads. Thee Candles Ltd, Dublin
Ó Murchada D, Murray M (2000) Place-names. In: Buttimer N, Rynne C, Guerin H (eds) The heritage of Ireland: natural, man-made and cultural heritage: conservation and interpretation, business and administration. Collins, Cork, pp 146–155
O'Drisceoil D, O'Drisceoil D (2011) Serving a city. The story of Cork's English market. Collins, Cork
Ó'Cuív B (1983) Observations on the book of Lismore. Proc R Ir Acad 83C:269–292
O'Flaherty R (1844) A chorographical description of West or Iar–Connaught. Irish Archaeological Society, Dublin
O'Leary DP (1970) A low head elver trap. EIFAC Consultation on Eel Fishing and Technologies, Hamburg
O'Meara JJ (1951) The first version of the topography of Ireland by Giraldus Cambrensus. Dundalgan, Dundalk
O'Sullivan A (1998) The archaeology of lake settlement in Ireland. Royal Irish Academy, Dublin
O'Sullivan A (2003) A day in the life of a medieval fisherman … and of intertidal archaeologists. In: Fenwick J (ed) Lost and found. Wordwell, Wicklow, pp 223–245
O'Sullivan A (2007) Exploring past people's past interactions with wetland environments in Ireland. Proc R Ir Acad 107 C:147–203
O'Sullivan A, Breen C (2007) Maritime Ireland. An archaeology of coastal communities. Tempus Publishing, Stroud
Parsons J, Vickers KU, Warden Y (1977) Relationships between elver recruitment and changes in sex ratio of silver eel *Anguilla anguilla* L. migrating from Lough Neagh, Northern Ireland. J Fish Biol 10:210–229
Poole RW, Reynolds JD (1996) Growth rate and age at migration of *Anguilla anguilla*. J Fish Biol 48:633–642
Ronan SG (1997) Irish writing on Lafcadio Hearn and Japan. Global Oriental, Folkstone
Rosell R, Evans D, Allen M (2005) The eel fishery in Lough Neagh, Northern Ireland – an example of sustainable management? Fish Manag Ecol 12:377–385
Scott M (1991) River gods. Real Ireland Design, Bray
Shaw-Smith D (1984) Traditional crafts of Ireland. Thames and Hudson, London
Smith C (1774) The ancient and present state of the county and city of Cork, 2. W. Wilson, Dublin
Smythe D (1988) A guide to Irish mythology. Irish Academic Press, Dublin
Squire C (1905) Celtic myth and legends. Gresham Publishing, London
Stokes W (1893) The voyage of the Hui Corra. Revue Celtique 14:22–69
Tesch F-W (2003) The eel. Blackwell Science, Oxford
Thompson W (1856) The natural history of Ireland. Henry G. Bohn, London
Waddell J (2005) Foundation myths. The beginnings of Irish archaeology. Wordwell, Bray
Waddell J (2010) The prehistoric archaeology of Ireland. Wordwell, Dublin
Went AEJ (1944) The Galway fishery. An account of the modes of fishing together with notes on other matters connected with the fishery. Proc R Ir Acad 49C:187–220
Went AEJ (1945) Fishing weirs of the River Erne. J R Soc Antiq Irel 75:213–223
Went AEJ (1950) Eel fishing at Athlone: past and present. J R Soc Antiq Irel 80:146–154
Went AEJ (1951) Fishing scenes from Irish topographical prints. J R Soc Antiq Irel 81:156–160
Went AEJ (1959) Sprat or white-fish weirs in Waterford Harbour. J R Soc Antiq Irel 89:91–93
Went AEJ (1964) The pursuit of salmon. Proc R Ir Acad 63C:191–244

Went AEJ, Mitchell NC (1952) Irish fishing spears. J R Soc Antiq Irel 82:109–134

Went AEJ, Mitchell NC (1956) Historical notes on the fisheries of the River Suir. J R Soc Antiq Irel 86:192–202

Westropp TJ (1892) Killaloe: its ancient palaces and cathedral. 1. J R Soc Antiq Ir 2(4):398–410, 5th Series

Wilson TG (1943) Some Irish folklore remedies for diseases of the ear, nose and throat. Ir J Med Sci 18:180–184

Woodman PC (1978) The chronology and economy of the Irish mesolithic: some working hypotheses. In: Mellars P (ed) The early post-glacial settlement of Northern Europe. Duckworth, London, pp 333–369

Yokouchi K, Aoyama J, Miller MJ, McCarthy TK, Tsukamoto K (2009) Depth distribution and biological characteristics of the European eel *Anguilla anguilla* in Lough Ennell, Ireland. J Fish Biol 74: 857–871

Eels in Culture, Fisheries and Science in Denmark

3

Suzanne Rindom, Jonna Tomkiewicz, Peter Munk, Kim Aarestrup, Thomas Damm Als, Michael Ingemann Pedersen, Christian Graver, and Carina Anderberg

As in many other parts of the world, eels and their mysterious life cycle have always fascinated Danes. Almost everyone in Denmark, no matter their age, knows something about eels. In fact, the eel was once one of the country's most important food fish, and Denmark itself was one of the main European nations fishing it, partly because of the seasonal abundance of migrating silver eels leaving the Baltic Sea through the narrow Straits of Denmark. Eels were fished year-round even during winter (Fig. 3.1). Although the Danish eel fishery was carried out mainly by smallholders, eels were for many years traded extensively with other European countries. Today, though, fisheries for eels are limited by low abundance and consequently restrictive laws.

Denmark has a long tradition of research into the biology and ecology of the European eel (*Anguilla anguilla*). From the early 1900s, when Danish scientist Johannes Schmidt identified its spawning area in the Sargasso Sea, right up to today, scientists have actively researched eel migration and larval drift. Also, the laboratory studies on reproduction and spawning behaviour of eels by Inge and Jan Boëtius during the 1960s and 1970s raised interest in research on eel physiology and behaviour. Today, Danish research on the genetics, migration and captive reproduction of eels is widely respected.

Many aspects of eel biology still remain a mystery, however, so Danish research on European eels continues in a national as well as an international context. Some of the history and current knowledge is outlined briefly here, and we also discuss the long-standing relationship between humans and eels in Denmark, including looking at the significance of eels in Danish culture and society, e.g. in fisheries and as food.

Eel Biology

Like all species of freshwater eel (*Anguilla* sp.), the European eel is catadromous, i.e. they spawn in the sea and grow in freshwater or brackish habitats. Although spawning eels and their eggs have yet to be captured in the wild, the presence of newly hatched larvae in the Sargasso Sea in the western Atlantic reveals that to be the spawning area of the species. Those yolksac larvae develop into leaf-like, transparent leptocephalus larvae that drift with the ocean currents towards Europe and North Africa. On encountering the continental shelf, they transform into transparent glass eels, which have the characteristic eel shape and migrate actively into coastal waters; in Denmark, glass eels appear at the coast, usually from April to July. Then, while developing into pigmented elvers, they disperse into estuaries, fjords, river mouths, rivers and lakes and assume the form of yellow eels, characterised by a dark upper body and a yellowish abdomen. They feed on a variety of food, such as worms, caterpillars, grubs, shrimps, frogs, fish eggs, fry and small fish, their growth rate varying with temperature and food availability. In cold, fast-running streams, for instance, growth is slower than in shallow lakes, where the temperature can be high in summer. On average, eels in Denmark grow 2–5 cm year^{-1}. Sex is determined early in the yellow eel stage and sex ratios have been suggested to vary mainly

S. Rindom (✉)
National Institute for Aquatic Resources
(DTU Aqua), Technical University of Denmark,
Jægersborg Allé 1, 2920 Charlottenlund, Denmark

National Food Institute, Technical University of Denmark,
Mørkhøj Bygade 19, 2860 Søborg, Denmark
e-mail: suri@food.dtu.dk

J. Tomkiewicz • P. Munk • K. Aarestrup • T.D. Als
• M.I. Pedersen • C. Anderberg
National Institute for Aquatic Resources
(DTU Aqua), Technical University of Denmark,
Jægersborg Allé 1, 2920 Charlottenlund, Denmark

C. Graver
The Danish Eel Farmer Association,
Agertoften 16, 6760, Ribe, Denmark

K. Tsukamoto and M. Kuroki (eds.), *Eels and Humans*, Humanity and the Sea,
DOI 10.1007/978-4-431-54529-3_3, © Springer Japan 2014

Fig. 3.1 Winter eelers on the island of Lyø, Denmark, in the mid-1950s (photograph, Ole Højrup, © Thomas Højrup. In Fisherman and Boats on Roskilde Fiord: Past and Present by Rieck Flemming and Max Vinner, 1989. The Viking Ship Museum, Roskilde, Denmark)

as a result of differences in population density; in bays and lagoons where densities are high, males tend to dominate, whereas in areas of lower density, such as freshwater lakes and the brackish Baltic Sea, it is females that tend to dominate. As a rule, yellow eels tend to be more common in shallow brackish water along protected coasts, and in fjords and bays, than in freshwater or the open sea.

Eels at most of their life stages seem to be negatively phototrophic, preferring darker habitats. They have a remarkably acute sense of smell, their specialised olfactory organ being five times the size of other freshwater fish and used to find prey, mainly at night. Another notable characteristic is their ability to survive more than 24 h out of water, allowing them to travel over land through wet vegetation when necessary, for example when migrating from isolated pockets of freshwater into rivers during the early stages of their spawning migration.

After some 6–20 years in fresh or brackish water, but depending on sex and physical state, the silvering process starts, indicating the onset of sexual maturation. During this process the eels adopt the silver form, with enlarged eyes and a light metallic abdominal colour, and they cease feeding. The main spawning migration takes place from late summer to early winter in Denmark, but migrating eels may be encountered too in spring. Moon phases seem to play an important role in the migration, but factors such as intensity of water flow are apparently equally important. During the silvering process, gonad maturation is inhibited, and maturation seems to be resumed only when approaching the Sargasso Sea. Eels are assumed to die after spawning because adult eels have never been observed in the Sargasso Sea or returning across the Atlantic.

Eel Habitat and Distribution in Denmark

Denmark is a small, flat country, with an area of ~43,000 km^2, and its highest point is just 173 m above sea level. Its coastline including all islands and fjords exceeds 7,300 km, and no place is more than 50 km from the nearest coast (Fig. 3.2). As there are no natural barriers to eel migration in Denmark, eels are found in virtually all freshwater rivers, streams, ponds and lakes as well as in brackish water along the coastline and in fjords. However, the Danish mainland and its islands are intensively cultivated or in other ways influenced by human activity, which has reduced natural eel habitats in both number and quality.

Wetlands have been drained and watercourses straightened and regulated, often with constructions hampering eel migration. Lakes, fjords and shallow coastal areas are subject to nutrient run-off from cultivated soils via freshwater streams, which through eutrophication may cause oxygen depletion in warm summers. Eels are sensitive to such oxygen depletion, although they may survive anoxic conditions for a short time. Efforts are now ongoing in Denmark as elsewhere to restore habitats and to reduce nutrient run-off to the aquatic environment, in support of recent management plans required by the European Union (EU) to aid recovery of the European eel.

The Status of Eels and Their Management

The European eel has declined significantly and is at a historical low throughout its distribution. Its decline is believed to have been caused by a combination of fishing pressure at all continental life stages (glass, yellow and silver eels), deterioration of habitats, including drainage of wetlands, the installation of hydro-electric plants and other barriers to up- and downstream river migration, pollution and introduced diseases and parasites. Eels were abundant in Danish waters until some 50 years ago, as much as 3,000–5,000 t of yellow and silver eels being taken annually between 1920 and 1970; by 2010, however, the annual catch was <400 t (Fig. 3.3). Seasonal runs of glass eels to Denmark are now very small and the recruitment of yellow eels to Danish streams is <5 % of that in the 1970s (ICES 2009a). Based on the state of the stock Europe-wide, the International Council for the Exploration of the Sea (ICES) recommends that "all anthropogenic impacts on production and escapement of eels should be reduced to as close to zero as possible until stock recovery is achieved" (ICES 2009b), and the European eel is now included in both the International Union for Conservation of Nature (IUCN) Red List of Threatened Species (Freyhof and Kottelat 2010) and the Danish Red List as "Critically Endangered" (Wind and Pihl 2004).

Fig. 3.2 A map of Denmark, showing its many fjords, bays and islands, which together create a coastline >7,300 km long

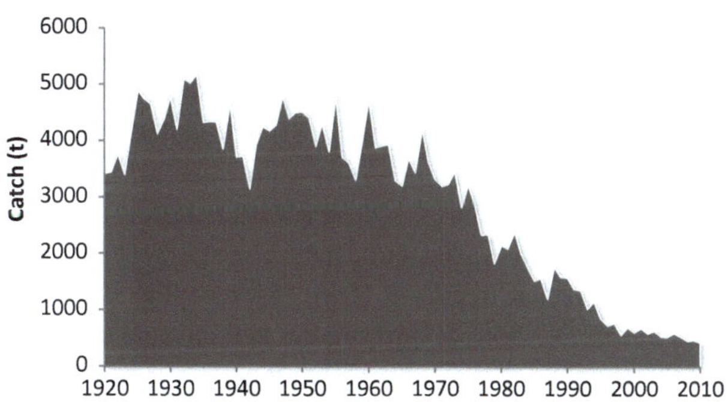

Fig. 3.3 Danish eel catches, 1920–2010 (statistics courtesy Danish AgriFish Agency, Danish Ministry of Food, Agriculture and Fisheries)

In 2007 a framework regulation for the recovery of the European eel was endorsed by the EU (European Union 2007), aimed at protecting and improving the possibility of future sustainable exploitation of the species. In terms of that framework agreement, each EU Member State is required to establish eel management plans for recovering the European eel population. The EU Council Regulation was implemented in Denmark in 2009 (Anon 2008). The Danish eel management plan consists of two primary elements: a plan for inland freshwater aiming at a 40 % escapement level (relative to pristine) for silver eels to the sea, and a management plan for marine waters reducing the fishing effort by at least 50 % relative to the average annual effort between 2004 and 2006. The measures proposed to attain these goals include a licensing system for professional fisheries and a shorter fishing season for recreational fishers. In freshwater, the measures further include restocking based on farmed glass eels, and in marine waters, an increase in the minimum legal capture size for yellow eels to 40 cm. Other measures in the plan include restrictions on fishing gear and season, catch registration, consideration of obstructions to migration, improvement of eel habitats, and restocking. Current adjustments to the measures are part of the plan, and depending on stock development trends, all inland eel fisheries may be phased out by the end of 2013.

Eel Fisheries in Denmark

The eel has been fished as long as there have been people in Denmark, and for good reason. The species was, at least until 50 years ago, abundant and distributed widely; it is fat and tasty, can be kept alive easily, is caught with simple fishing gear, is available year-round, and has a predictable migration pattern. There is archaeological evidence in Denmark that eels were being captured by pot (Fig. 3.4) and spear and used as a source of food in the Stone Age (~5000–4000 BC).

Up to World War 2, fishing for eels in Denmark was typically performed by smallholder farmers, who fished for eels to supply the household with food or extra income. After World War 2, however, the practice changed and professional fishing for eels gained popularity with the use of trawls, nylon nets and larger poundnets; catches were sometimes huge. However, as the stock and the catches dwindled, so did the number of commercial fishers, and nowadays only a few are involved in professional eel fisheries (Sparrevohn et al. 2011). Many more recreational fishers catch eels in Denmark, though, perhaps as many as 15,000, but the types of gear that they can use are restricted and purchase of their catches is illegal. Recent surveys estimated that recreational fishers may catch some 100 t of eels annually (Sparrevohn et al. 2011), an amount not accounted for in the annual 400 t commercial catch.

Most Danish eel fisheries, commercial and recreational, operate in the brackish coastal waters of shallow fjords and bays, where eel density is greatest, and just 5 % of the catches reported are derived from freshwater. The Danish fishery targets yellow and silver eels, with fisheries on yellow eels between spring and early autumn and on silver eels between late summer and late autumn. The main fishing seasons for yellow and silver eels depend, however, on factors such as temperature, weather, water flow and discharge. Most commercial eel fisheries are concentrated in southern and eastern Denmark, where silver eels pass through the narrow Straits of Denmark as they migrate from the Baltic Sea towards the North Sea and Atlantic, and ultimately the spawning area in the Sargasso Sea.

Fig. 3.4 A fish pot made of woven osiers from the early Stone Age. The pot, once used for eel fishing, was found in a bog near the town of Holbaek in Zealand, Denmark, and is today exhibited at the National Museum of Denmark. This type of fishing gear has been used almost unaltered until today, either alone or as part of an eel trap (from www.denstoredanske.dk; Danish open encyclopaedia)

Traditional Fishing Gear and Methods

A wealth of methods and equipment has been developed and used for fishing eels. In Denmark, the gears used through time include a variety of spear types, eel pots and traps, rod and line, longlines, pound-, fyke- and seinenets, and trawls. The use of some gear types, e.g. spears (Fig. 3.5), stopped long ago, and other types were prohibited more recently as part of the regulation of fisheries. Since 2009, for instance, the use of driftnets, trawls and seines has been prohibited and seasonal closures have been implemented for all remaining fisheries. Recreational fishing has been strictly limited in terms of gear and operating season.

An *eel pot* is an ancient fishing gear that has persisted almost unaltered over time, but it is rarely used today. The oldest known eel baskets found in Denmark were woven from osier (Fig. 3.4); today they are made of synthetic fibre or plastic. An eel pot can be used alone or in connection with a *fykenet* (Fig. 3.5 bottom left), individually deployed in streams or lakes. *Eel spears* (Fig. 3.5 top left) were used mainly from boats during summer, but winter eelers in the past would also catch eels through holes in the ice (Fig. 3.1). Spears were then thrust into the muddy bottom to catch the eels. Different types of spear were used depending on the nature of the bottom and the size of the eels being targeted. In Denmark, all types of spear are now prohibited, because they harm the eels and prevent the successful release of undersized fish.

A traditional way of catching eels in shallow water in the fjords and freshwater especially in northern Denmark was eel *pushnets* or *push traps* (Fig. 3.5 top right). To operate them, two people worked together, one working the bottom and disturbing the resting eels, chasing them into the pushnet, the other lifting the net as soon as an eel entered it. The catch was kept in a basket carried around on the shoulders of the operators. *Eel weirs* (Fig. 3.5 bottom right) are no longer used in Denmark, although they were until a few decades ago. A bridge on top of the leader made it convenient to harvest eels from the eel pot at its end. The legal right to fish with them belonged to the owner of the coastal land, but nowadays there are no private fishing rights at the coast. In some places there are wooden *eel traps* (Fig. 3.6), built at the outflow of lakes or on weirs in rivers to catch the silver eels migrating downstream during autumn. All such traps are registered and their use restricted by seasonal closure. From the end of 2013, however, their use will be prohibited completely.

Longlines may operate with 100 or more hooks baited with fresh worms, shrimps or small fish. Lines are set in the evening and checked before sunrise. The use of such longlines is still legal in some areas, but not in others. In earlier times, a simpler but not necessarily easy way to catch eels on a line was *bobbing* (peuren in the Netherlands), which required 10–20 earthworms to be coiled into a ball on a thin thread fastened to a string at the end of a rod. When the bait was lowered into the water, an eel would bite it and before it let go, would be lifted into the boat; great skill was required to fish in this way!

Poundnets are probably the most commonly used type of gear in recent time, especially in the straits connecting the Baltic with the North Sea, where they effectively catch migrating silver eels. When the fishery was most intense, poundnets were placed strategically just 500 m apart over many kilometres. For poundnets targeting silver eels (Fig. 3.7 top), which are legal only for professional fishers, the diameter of the enclosure is typically >20 m and the headline up to 500 m long. The latter is fixed on piles or attached to buoys. The nets themselves are constructed to fit individual fishing sites and can be quite large, sometimes extending from the surface to depths of 16 m. *Fykenets* (Fig. 3.5 bottom left and Fig. 3.7 bottom) are smaller and have no enclosure, and the headline is typically <50 m long. Their headline too is fixed on piles or attached to buoys. Both pound- and fykenets are still operated all around Denmark, and in fjords and lakes, most often in mixed fisheries, but there are tight restrictions on their use. The use of *driftnets* (Fig. 3.8), *seines* (which can also be deployed from shore) and *trawls* to catch eels is now prohibited in Denmark.

Fishing Boats

Eel drifters were developed and built from 1880 to 1930 exclusively to fish for eels (Fig. 3.8). They were classic Danish, clinker-built boats, constructed so that they could drift sideways with an extended seinenet drifting out from their side. When the centreboard was lifted, the flat-bottomed vessel

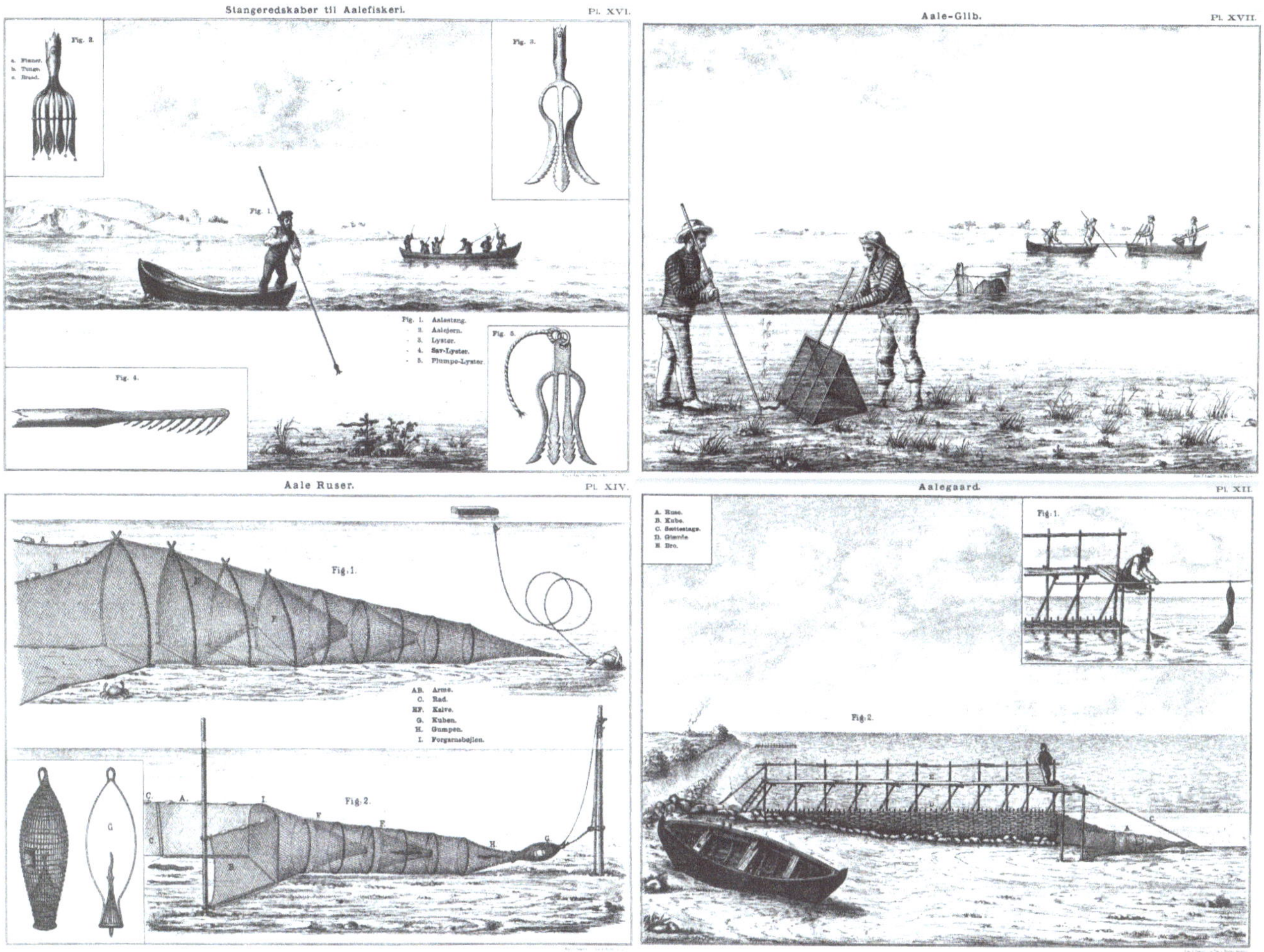

Fig. 3.5 Early types of gear used to fish eels: eel spears (*top left*), eel pushnet (*top right*), a selection of fykenets (*bottom left*) and an eel weir (*bottom right*) (from Drechsel 1890)

Fig. 3.6 A freshwater eel trap at the stream leading from Lake Vandet in Thy National Park, Denmark, used to catch migrating silver eels (photograph, Sune Riis Sørensen, DTU Aqua)

would simply drift sideways, dragging the net along the seabed. The catch was kept alive in a well amidships. However, with the introduction of motorized boats, such drifters were consigned to history as far as fishing was concerned. Some 10 eel drifters still exist, though, and they are used for recreation because of their fine sailing abilities (Fig. 3.9).

Restocking of Eels

During industrial and agricultural development in the eighteenth and nineteenth centuries, many river systems were altered by dam construction for hydro-electric power production or weir building to control water levels, and both created barriers to eel migration. To mitigate the effect of such barriers, however, it has been statutory in Denmark since 1898 to install eel passes at the barriers. In the past, elvers and yellow eels were stocked by fishers in inland waters, in places where recruitment was low; indeed, from the mid-1960s to the end of the 1980s, licences were granted to catch and sell eels for stocking purposes. Such stocking and transfer of eels was then prohibited as a consequence of the poor recruitment in the wild and to prevent the spread of

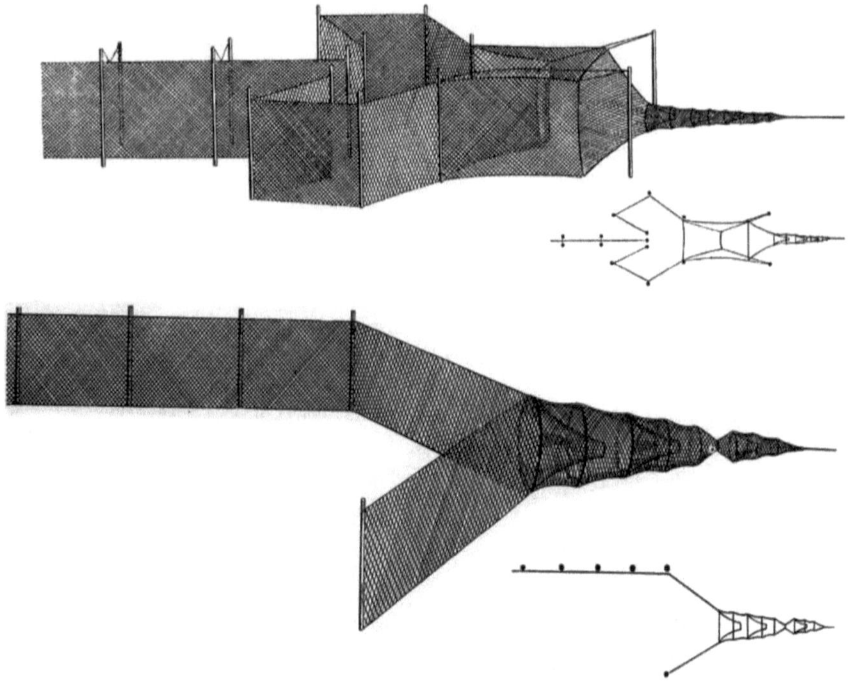

Fig. 3.7 A typical present day poundnet (*top*) and fykenet (*bottom*) used for catching eels (from Anon 2008)

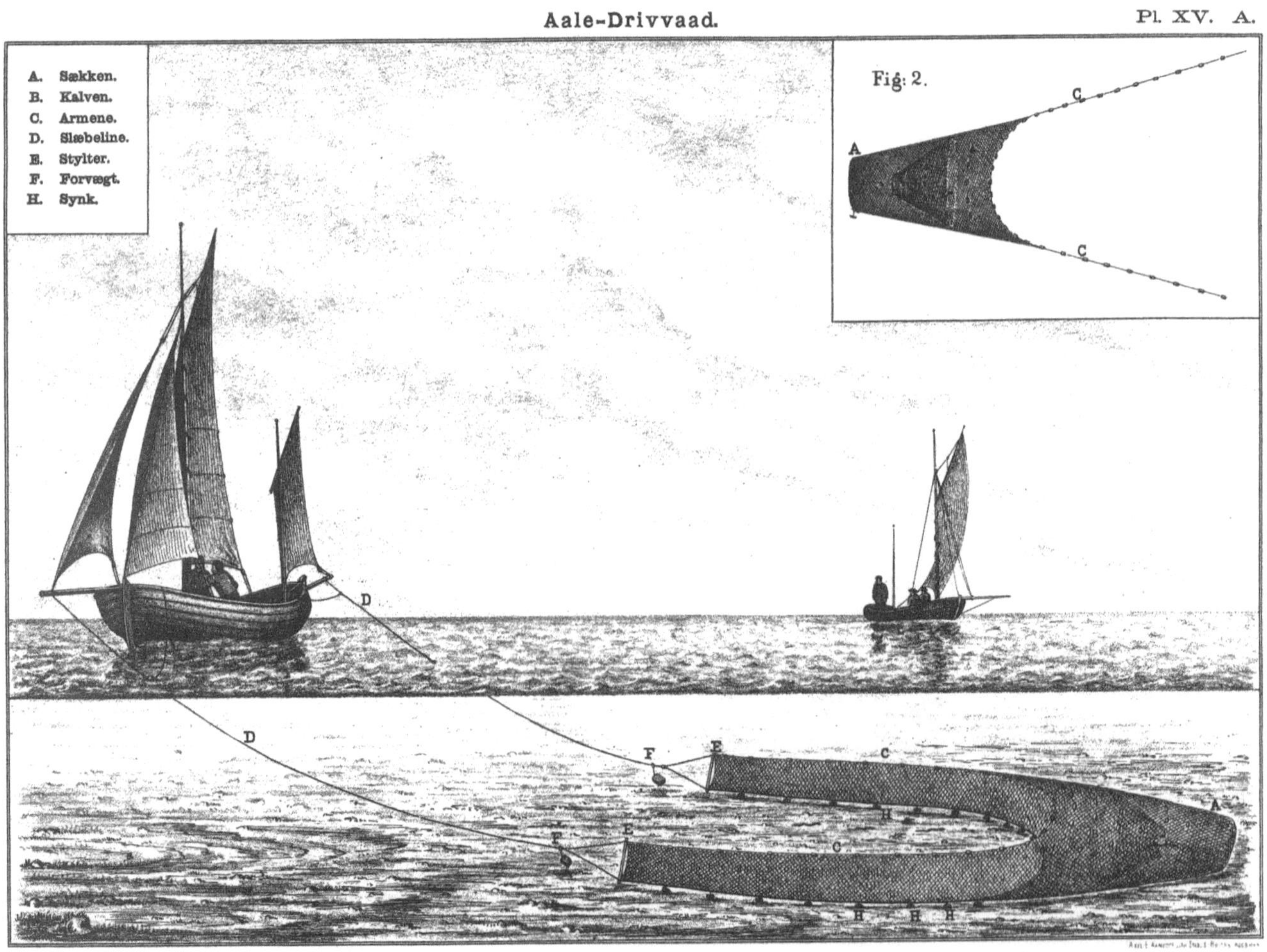

Fig. 3.8 An eel drifter with a driftnet (from Drechsel 1890)

Fig. 3.9 Eel drifters now used for recreational sailing: "Concordia", in private ownership (*left* and *centre*) and "Viktoria" owned by the Viking Ship Museum (*right*). (Photographs left and centre, Suzanne Rindom, DTU Aqua; right, Werner Karrasch, © The Viking Ship Museum, Roskilde, Denmark)

the eel swimbladder worm *Anguillicoloides crassus*, which was introduced accidentally to Europe from Asia during the 1980s. Since 1987, however, the Danish Government and the recreational fishing licence fee have financially supported a controlled restocking programme (Kirkegaard 2010), using wild-caught glass eels imported mainly from France and grown to a weight of 2–5 g (fingerlings) before release. Recently, the quantity stocked has declined, following the escalation in the price of glass eels. In 2010, 1.54 million fingerlings were stocked in Danish lakes, rivers and coastal areas.

Eels in Aquaculture in Denmark

Eels have been farmed in Denmark for centuries. Initially, there was no distinction between restocking and farming, and eels were moved around as glass eels, fingerlings and elvers, often using ponds for farming purposes. The eels were fed many types of feed, but mainly fish waste. A problem with farming in ponds, however, is the eel's ability to escape over land through wet grass, hence at times leaving farmers with no eels to harvest!

In the late 1970s, first attempts were made to farm eels in indoor aquaculture facilities using recirculation technology (Kirkegaard 2010). Such facilities were often established in existing buildings and consisted of fish tanks, a reservoir, a pump and submerged filters (which allowed bacteria to remove waste products from fish by nitrification of the ammonia), supplemented with an oxygenating device. Many such systems were built during the 1980s, but they required a huge amount of water, as much as 1,000 l kg^{-1} of feed. Also, the water had to be heated to ~25 °C, which even in those days was financially challenging. Later, therefore, microsieves were introduced to remove particles prior to cleansing by the submerged filters, and ultraviolet light was employed to sterilize the water. A modern-day purpose-built eel aquaculture facility and a filtration plant are shown in Fig. 3.10. The main reason for the huge quantity of water needed in the process was the nitrate created by nitrification at the submerged filters, so farms started to use denitrification technology too. Nowadays, with this technology, <100 l of water are needed per kg of feed, and some farms have also introduced phosphorus systems, reducing the water requirement to <50 l kg^{-1} of feed.

Glass Eels and Farming

Aquaculture eel production is currently based exclusively on wild-caught glass eels, which for the European eel are fished in estuaries and rivers mainly in the Bay of Biscay and the UK's River Severn (Kirkegaard 2010). Like everywhere else, though, catches of glass eels there have decreased dramatically over the past 50 years, in France for instance dropping from >500 t in 1995 to ~50 t in 2011. During those 15 years too, Chinese eel culturists have shown great interest in purchasing European glass eels, because Japanese glass eels (*Anguilla japonica*) tend to be up to ten times more expensive than European ones.

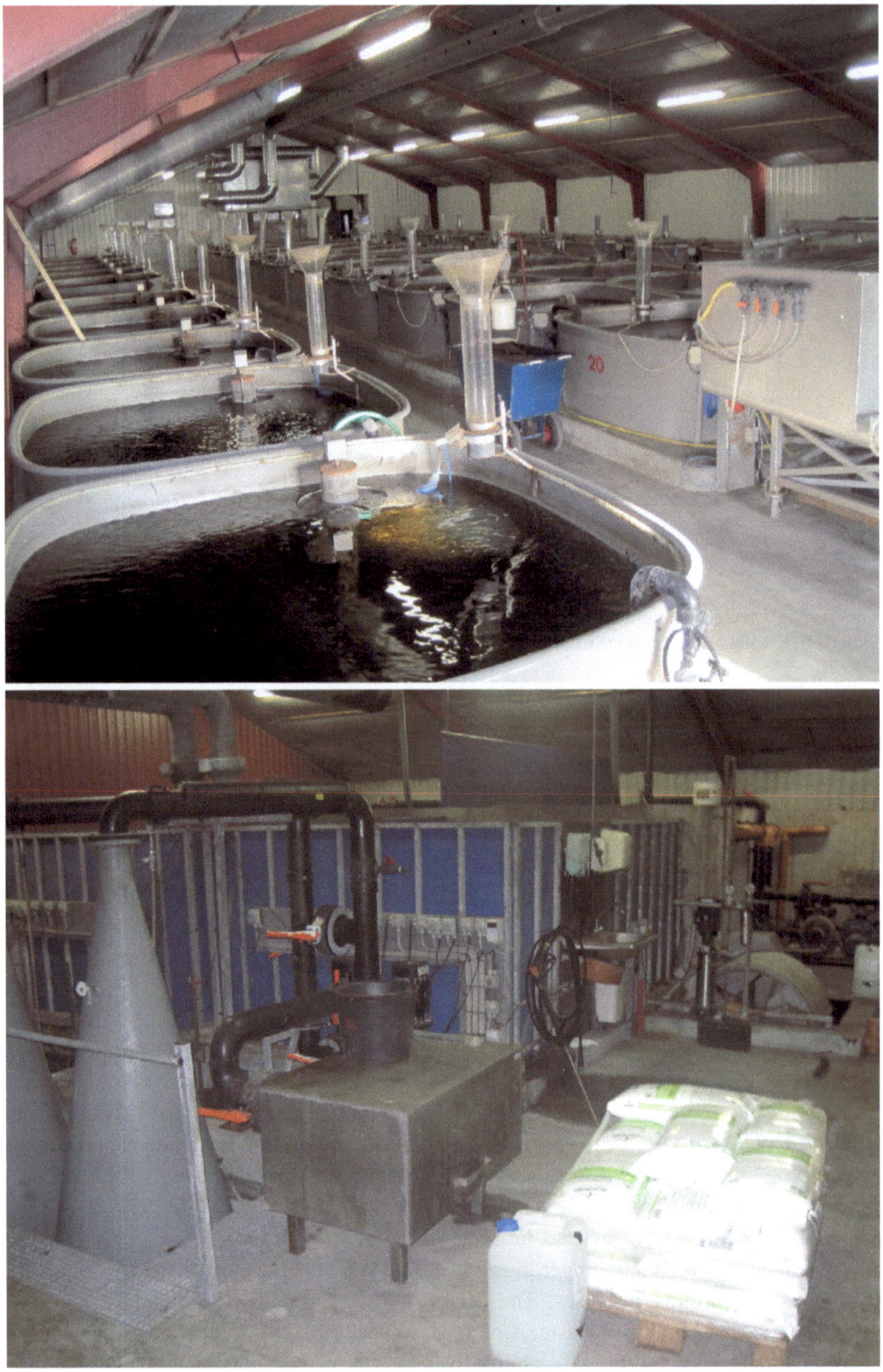

Fig. 3.10 Fish tanks at a Danish eel farm (*top*), and a filter unit with submerged and trickling filters, oxygen cones, pumps, UV-sterilizer, microsieve and pumps (photographs, Christian Graver, Danish Eel Farmers Association)

Increased competition has therefore driven up the price of European glass eels from €200 to as much as €1,100 per kg, although prices are generally fairly stable at ~€400–600 per kg. Trading restrictions are now implemented to protect the European eel.

The continuing decline in the catch of glass eels naturally stimulated the various research initiatives underway in Europe and elsewhere in the world to develop a self-sustained culture of eels, focusing on breeding programmes for aquaculture. This subject is addressed also elsewhere in this book.

Eel Production, Markets and Restocking

During the 1980s, most Danish aquaculture production came from small, family-owned businesses delivering between 1 and 10 t annually. Then, during the 1990s and into the twenty-first century, the number of eel farms decreased, and the average annual production of each of the remaining businesses grew. In 1990, for instance, there were 47 eel farms in Denmark with an average annual production of 12.5 t; by 2011 the number had dwindled to just eight, however, with an average annual production of ~250 t.

Most of the eels produced in Denmark for human consumption are exported to the Netherlands. Dutch smokehouses prefer eels of ~145 g for smoking and filleting; eels >400 g are smoked in Denmark or exported to Germany. A substantial proportion of glass eels entering current aquaculture in Denmark is used in the production of fingerlings to restock Danish and other northern European water bodies; this includes several million fingerlings of 2–10 g that are exported to Germany and Poland.

Danish Eel Research

Denmark has a long tradition of research into the biology and ecology of the European eel. Past and present research areas include the identification of spawning areas, silver eel migration and analyses of the factors that influence larval drift and development, along with population genetics and studies on the reproduction of eels in captivity.

Johannes Schmidt's Research

Johannes Schmidt (see box) is renowned for his research during the period 1904–1922 outlining the spawning area of the European eel. His great interest in finding the spawning areas and larval drift routes was aroused when he caught a leptocephalus larva off the Faroe Islands in 1904. Eight years before, in 1896, the Italian zoologist Giovanni Grassi had been the first to identify eel larvae by observing the transformation of a leptocephalus (formerly believed to be a unique species) into a glass eel (Grassi 1896), and suggested that spawning took place in deep waters of the Mediterranean Sea. However, on finding eel larvae in the Atlantic, Schmidt argued that at least some spawning areas would be in the Atlantic itself. Dedicated to resolving the enigma of eel reproduction, he initiated a Danish research programme that focused on identifying the spawning areas of European eel, and because of the importance of the eel to Denmark, he attracted financial support for a campaign that eventually lasted for two decades, interrupted only by World War 1.

In the early 1900s, researchers had succeeded in documenting the spawning areas, stock distributions and spawning migrations of various fish species. During his early work on other species, notably Atlantic cod (*Gadus morhua*), Schmidt had worked with methods for identifying fish spawning areas, including the development of gear that effectively sampled fish eggs and larvae, so he systematically sampled eel larvae over large areas of the Atlantic, obtaining information on species composition and larval length. Then, by back-tracking larval size to the locations where the smallest larvae were caught, he mapped drift patterns and delimited the spawning area of the European eel as the Sargasso Sea.

Schmidt attempted to publish his initial results in 1912, but he was not acknowledged for his findings at that time, because confirmation of the delimitation of the exact area of spawning was required. In fact, he was refused the opportunity to publish his results in the Proceedings of the Royal Society, the obvious scientific medium then for publishing results of such importance. He carried out a cruise in 1913 to the western Atlantic, but World War 1 came and 7 years passed before he had another opportunity to return to the Sargasso Sea. That survey took place in 1920/1921 on a schooner placed at his disposal by a privately owned trading company, the East Asiatic Company of Copenhagen. During the survey he sampled larvae in the area he had earlier identified to be the spawning area of both American (*Anguilla rostrata*) and European eels, and he then delimited the area by collating the information obtained during this and preceding surveys. Finally, in 1922, he published the work on eel larval abundance and size across the entire North Atlantic (Fig. 3.11) for which he later became famous (Schmidt 1922); in this he defined the spawning area of both European and American eels as the southern Sargasso Sea, based on a relatively narrow band of very small larvae centred around latitude 26°N. Another important result from his studies was a description of the increase in size and development of the growing leptocephalus larvae over time as they drifted from the spawning site towards the European continent.

Danish Field Research, 1966–2007

Danish interest in eel research was maintained after Schmidt's death in 1933, and his material, supplemented with new material from the eastern Atlantic, was analysed further. However, it was not until 1966 that another Danish field expedition set out for the Sargasso Sea with focus on the eastern part of the hitherto identified spawning area of the European eel, where small eel larvae had been found as far east as ~57°W.

During the 1970s and 1980s, several countries carried out research cruises to the Sargasso Sea, collecting new information on the distribution of eel larvae, and it became clear that

Professor Doctor Phil. Johannes Schmidt

Ernst Johannes Schmidt is one of Denmark's best known marine scientists and his work is respected worldwide, in particular his research on eels. He was born on 2 January 1877 (Bruun 1934), received a master's degree in natural history in 1898 and a doctoral degree in botany in 1903. Also in 1903, he married Ingeborg Kühle, daughter of the director of the Carlsberg Co. (Reagan 1933). During 1901 and 1902 he worked as an assistant at the Danish Biological Station, where his interest in marine science was stimulated by Prof. C. G. J. Petersen, then from 1902 to 1909 part-time for the Botanical Institute of the University of Copenhagen and part-time for the Danish Commission for Investigation of the Sea, on marine science. In 1910, he was made head of the department of physiology at the Carlsberg Laboratory, a post he held until his death on 21 February 1933 (Bruun 1934).

Johannes Schmidt (photo © Polfoto, Copenhagen, Denmark)

In his early research, Schmidt clarified aspects of the early life of commercial fish, especially cod. In 1903, however, he expanded his geographic horizons to cover the southern areas of the North Atlantic, where he collected large numbers of pelagic fish larvae around the Faroe Islands and Iceland. This material provided new, detailed knowledge about North Atlantic fish species during all their life stages, from eggs and larvae to adults, as well as information on spawning areas and seasons. It is, however, for his work on eels that Johannes Schmidt will be remembered best (Bruun 1934). After catching a single eel larva in the Atlantic in May 1904, he devoted the rest of his working life to solving the mystery of eel reproduction. His work ultimately led to the discovery that Atlantic eels breed in the Sargasso Sea, finally presenting proof of this in 1922 (Reagan 1933). Then, having demonstrated where Atlantic eels spawn, Schmidt turned his attention to other parts of the world. He travelled to Australia, New Zealand and Tahiti to study the eels of the Pacific, and it was those studies that fuelled his desire to map the spawning and larval stages of eels other than the Atlantic species. A large grant from the Carlsberg Foundation along with considerable state subsidies then supported him on a circumnavigation of the world in the steamer "Dana." The ship left Copenhagen on 14 June 1928 for the Mediterranean, crossed the Atlantic, passed through the Panama Canal, went past the Pacific Islands en route to New Zealand and Australia, and from there steamed on to China, New Guinea and Sumatra, across the Indian Ocean to Madagascar and East Africa, around the Cape of Good Hope, then back north through the Atlantic, arriving back in Copenhagen on 30 June 1930. Schmidt brought back large collections of marine fauna (Winge and Tåning 1947), which now reside in the Zoological Museum in Copenhagen, many open for public view. The Danish newspaper "Politiken" wrote under the headline *The* "Dana" *returns home after two-year voyage*[1]:

> Yesterday afternoon at about 2 pm, a small, grey steamer glided into Copenhagen harbour and set a course down past Langelinie. At the time there was a large welcoming committee awaiting it, at the Custom House. Reading the name "Dana" on the bows, those watching realized that it was the expedition that had set out 2 years before to investigate the riddles of the seven seas and which, under Professor Schmidt's leadership, had made Denmark's name known right round the world and boosted her reputation in the field of international ocean research.
>
> We managed to pull Professor Schmidt aside and asked him to tell us a little about the results of the voyage: "what do you consider to be the expedition's most important result?" "It will take a long time before that question can be answered", replied Prof. Schmidt. "Not until all the material has been analysed carefully will we be able to say for certain, but as far as I can tell at the moment, I believe that our research into the life and activities of the eel will be the most significant outcome".

[1]Selected passages from the Danish newspaper article "The Dana returns home after two-year voyage", issued in "Politiken" 1 July 1930, translated into English by Pete Westbrook, then paraphrased.

The "Dana" returns home after a 2-year voyage. Ocean research expedition brings back a huge amount of material that will take 5 years to collate (photo courtesy DTU Aqua, Denmark)

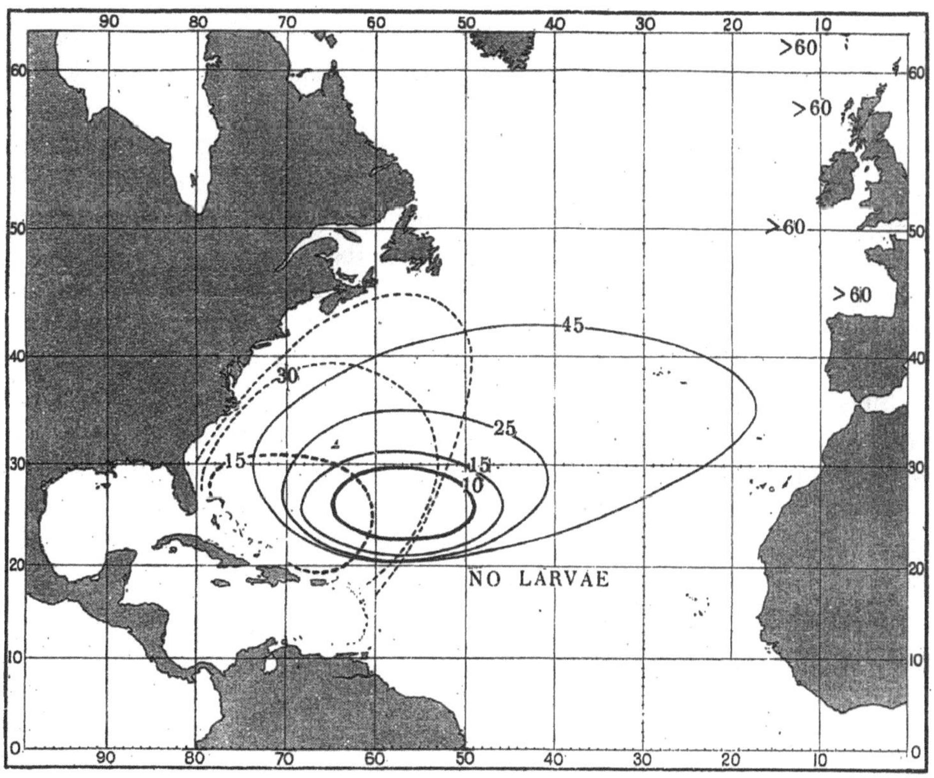

Fig. 3.11 European and American eel spawning areas, with the distributions of larvae of different size indicated by *curves* (*dotted* for American, *continuous lines* for European). The *heavily-drawn innermost curves* embrace the spawning areas of the two species, and the curves drawn progressively outwards from the spawning areas show the limits of occurrence of 25 mm specimens, etc. (from Schmidt 1922)

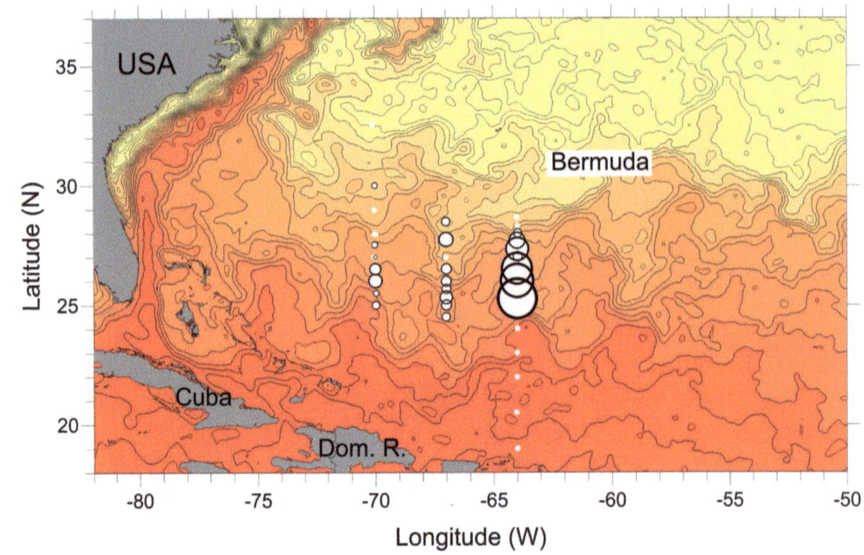

Fig. 3.12 *Top*: Satellite image of the Atlantic Ocean in April 2007, illustrating where warmer subtropical water (*red*) meets cooler northern water (*yellow*), indicating the frontal zones. The *white dots* illustrate the sampling for larvae in 2007, and *black circles* depict the abundance of European eel larvae (graphics courtesy Peter Munk, DTU Aqua). *Bottom:* The plankton net is hoisted on board the RV "Vædderen" during a haul in the Atlantic Ocean in 2007 (photograph, Peter Munk, DTU Aqua)

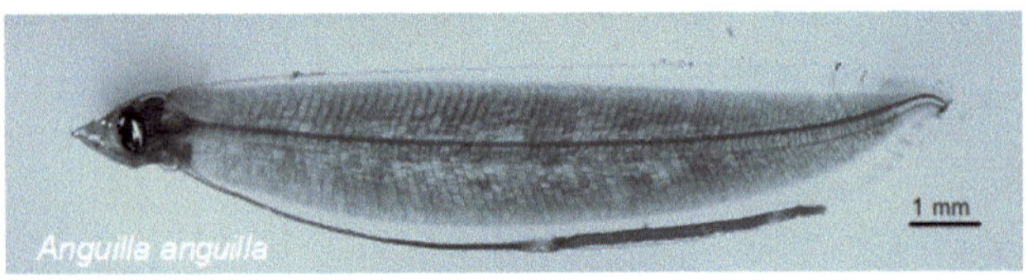

Fig. 3.13 A European eel larva (photograph, Peter Munk, DTU Aqua)

the distribution of small larvae, and consequently eel spawning, was related to the unique hydrography of that part of the Atlantic. There, warm tropical surface water meets cooler water from the North Atlantic and forms a front at the subtropical convergence, and it is there that the abundance of small eel larvae appears to be greatest (Fig. 3.12 top).

In 2007, a new investigation was carried out as part of the Danish Galathea 3 World Expedition. The research led to enhanced understanding of the location of eel spawning areas, larval life history and the potential influences of climate-induced environmental changes. During a cruise in March–April 2007, the RV "Vædderen" worked in the Sargasso Sea (Fig. 3.12 top). Transects crossed the fronts of the subtropical convergence with the aim of investigating hydrography, chemistry and plankton biology and intensively sampling for eel larvae (Fig. 3.12 bottom). Both American and European eel larvae were caught in the frontal zone, with sizes as small as 5 mm, though the average size was ~12 mm (Fig. 3.13). The findings underscored the importance of oceanic frontal processes in retaining eel larvae within areas of good food availability and in directing their drift towards continental land masses (Munk et al. 2010). As Schmidt had shown earlier, American eel larvae dominated in the western part of the area investigated and

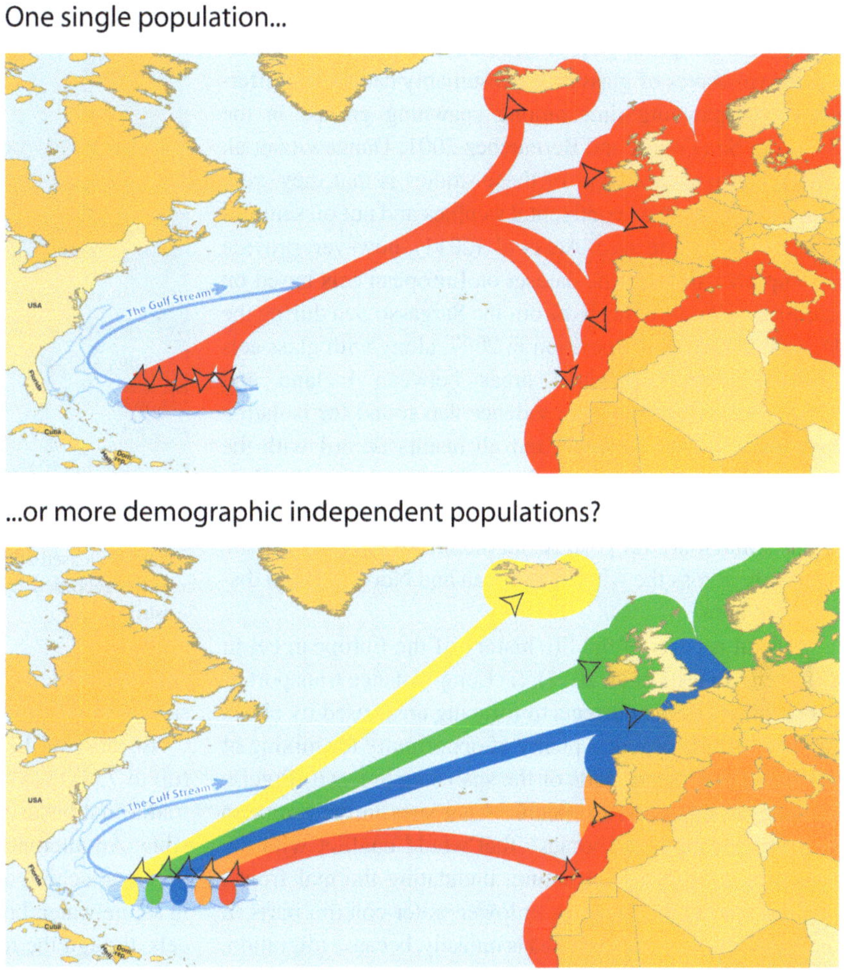

Fig. 3.14 An illustration of the two distribution theories for the European eel, panmixia (*top*) and genetic differentiation between populations (*bottom*). Graphics courtesy Thomas Dam Als, DTU Aqua

European eel larvae towards the east. Hence, it is likely that most of the American eel larvae will drift west or north in the Antilles Current, and that European eel larvae will tend to follow an eastward route toward the Azores and Europe, because of their more easterly distribution close to the east-flowing current associated with the front (Munk et al. 2010). Clearly, therefore, the life history, survival and drift of European and American eel larvae will depend strongly on climate-related oceanographic processes.

The European Eel: One or More Populations?

Another question central to the management and conservation of Atlantic eels is whether there is a single population or several smaller, demographically independent ones.

European and American eels appear to spawn in partially overlapping areas in the Sargasso Sea, with larvae of the American eel drifting ~2,000 km to North America and their European equivalents drifting >5,000 km to the coasts of Europe and North Africa. Several aspects of eel biology, including their long-distance spawning migration and subsequent lengthy larval transport by ocean currents, suggest that all individuals of each species target the same spawning area and mate independent of origin, despite such random mating being rare in geographically widespread species. Hence, both European and American eels represent classic textbook examples of panmictic species, each comprising a single, randomly mating population.

Schmidt separated European and American eels on the basis of the number of vertebrae, and he observed no geographic differentiation within either species, supporting the panmixia hypothesis (Schmidt 1922). More recently, the results of genetic studies of populations of the European eel (Fig. 3.14) have fuelled controversy. Early genetic studies supported panmixia in both species, but some later studies using novel genetic markers yielded evidence against panmixia, by suggesting isolation of populations by either migration distance or time. One set of studies found significant genetic differentiation between geographically distinct samples of glass eels, but another that indicated isolation by time found temporal genetic differentiation between different arrival waves of glass eels, presumably caused by differences in spawning time among spawning groups in the Sargasso Sea (Wirth and Bernatchez 2001; Dannewitz et al. 2005). Common for most of these studies is that they were based exclusively on continental samples and not on samples from the spawning area. Als et al. (2011), however, provide a comprehensive genetic dataset on European eels based on samples of larvae collected from the Sargasso Sea during the Danish Galathea 3 Expedition in 2007, along with glass eels sampled from continental areas between Iceland and Morocco. In that study, no evidence was found for isolation by either distance or time, and all results accord with the panmixia hypothesis, so on the basis of those results, it is likely that all silver eels of a species share a common spawning area and that European eel larvae subsequently distribute randomly across the whole European and North African distribution area.

Several aspects of the life history of the European eel fit this scenario (Als et al. 2011). (i) Long-distance transport of eel larvae by ocean currents to foraging areas used by elvers and yellow eels provides plenty of opportunity for mixing of larvae from different parts of the spawning area. (ii) Precise homing of silver eels to specific natal sites in the Sargasso Sea appears unlikely, because that would conflict with the necessity to spawn in dynamic, undulating thermal fronts. (iii) Imprinting and homing to lower-water-column parts of the 5,000-m-deep Sargasso Sea is unlikely, because migration, spawning and larval drift takes place in the upper pelagic zone of the ocean (<1,000 m deep).

Hence, the study concludes that European eels likely do not exhibit fine-scale homing towards their precise natal sites, but rather a crude homing behaviour towards the frontal zones of the Sargasso Sea (Als et al. 2011). The finding that all European eels form a single population points to the need for management and conservation effort to be coordinated at a transnational level; overexploitation in one region would negatively influence the whole population of European eels and depress recruitment across the entire distributional range of the species.

European Eel Migration: Following Silver Eels to Their Spawning Area

One of the main mysteries about eels is what happens when the silver eels leave the coasts of Europe never to return. Some eels have been caught incidentally in trawls south of the Faroe Islands and off the Azores, and a few have been identified in the stomach contents of humpback whales. Knowledge of the migration behaviour or route of the eels is still incomplete, however, and no-one has yet caught a spawning eel in the Sargasso Sea.

Fig. 3.15 A female silver eel with a pop-up satellite archival tag (PSAT), released at the west coast of Ireland after tagging (image from a video recorded by Robert Schabetsberger and Ingo Eichelberger, Austria)

Because of the very long distances European eels migrate (up to 7,000 km) over possible water depths of >5,000 m, a traditional fishery approach to studying migration is not feasible. An alternative way to track migration is by telemetry, i.e. by attaching a tag to the animal. Radio tags do not work in saline water, but acoustic tags do and have been applied to eels, though the results are limited by the restricted range of signal transmission. Recently, however, modern Pop-up Satellite Archival Tags (PSATs) have proven successful in documenting eel migration (though see the discussion on the subject elsewhere in this book). PSATs are used widely to track the movements of bigger animals that live in the upper few hundred metres of the ocean and that typically travel great distances, which makes them difficult to study in any other way. Tags are mounted on the animal and store data such as temperature, depth and light level. At a predetermined time, the tags initiate a release mechanism and "pop off" the animal, float to the surface and relay their data to ARGOS satellites, and that information can then be downloaded via the ARGOS system.

During the Danish Galathea 3 Expedition in 2007, researchers attempted to determine the transoceanic migration routes and behaviour of migrating silver eels and to test the concept of PSAT tagging in a large-scale field programme (Aarestrup et al. 2009). In order to obtain large eels for tagging, almost 100,000 eels were captured. Of these, 22 large wild female silver eels were selected and tagged with a small PSAT mounted on their backs, with a specially developed attachment securing the PSAT to the eels, like backpacks (Fig. 3.15). All were released on the west coast of Ireland in

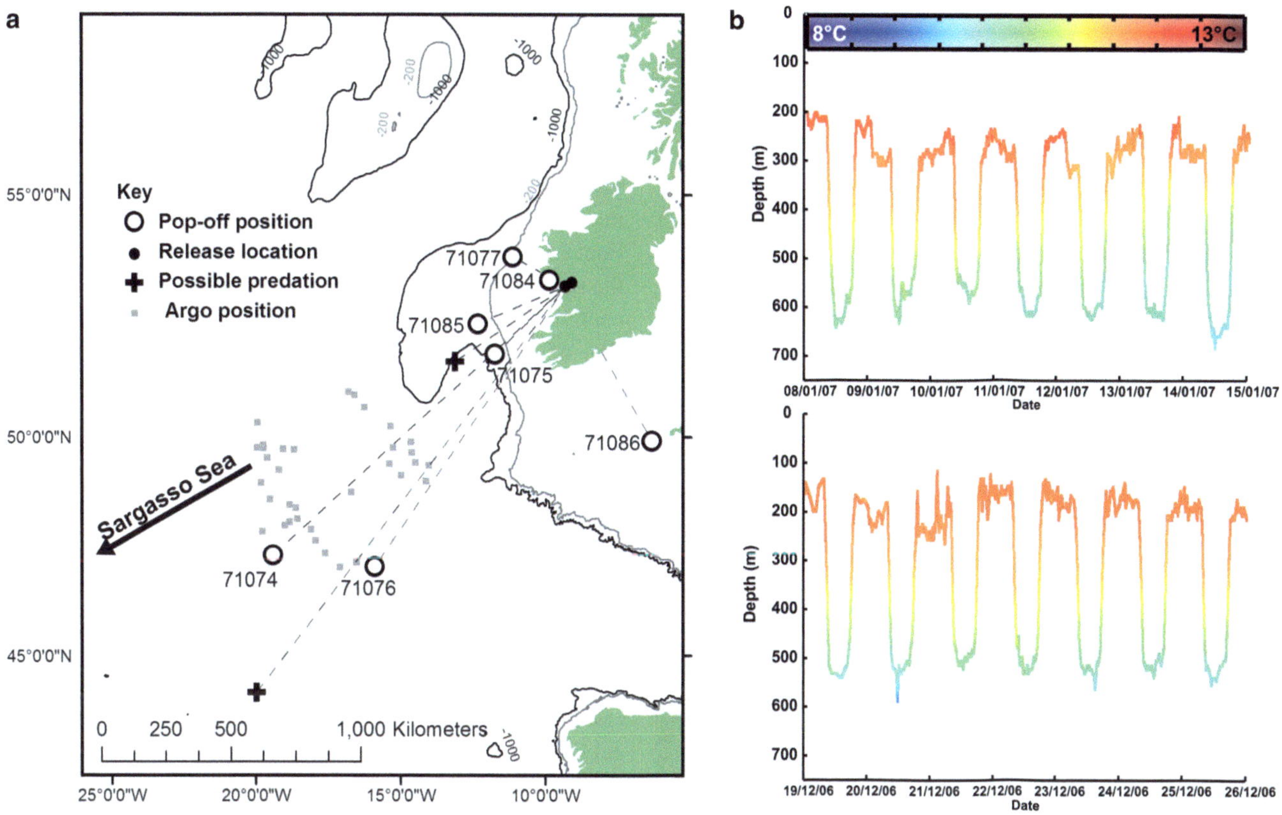

Fig. 3.16 (**a**) Map showing the pop-up positions for the PSATs. (**b**) and (**c**) Vertical migration of two eels over a week, with depth values coloured by temperature (from Aarestrup et al. 2009, reproduced with permission)

October and November 2007 during their normal migration season there. The PSATs were programmed to be released from the eel at a pre-set time, and although the full migration route was not completed, some tantalising insights were gained. The eels on leaving Ireland seemed to take a south-westerly route towards the Azores (Fig. 3.16a), south of the line they would have taken had they been heading directly for the Sargasso Sea, possibly to take advantage of favourable current systems and hence to save stored energy for reproduction. The study also demonstrated (Fig. 3.16b, c) that the silver eels migrated closer to the surface at night and deeper by day, once the eels had left the continental shelf and were in oceanic water often thousands of metres deep. Such a pattern, known as diel vertical migration, is not uncommon in marine organisms and is thought to be an adaptation evolved mainly as a trade-off between feeding and predation avoidance. By day, animals dive deep to avoid predators, but at night when predation risks should be lower, they rise towards the surface to feed in the more productive layers of the water column. The deep dives by day seem to make sense for eels, which obviously wish to lessen their risk of predation, but silver eels do not feed at all while migrating, so why they swim closer to the surface at night is not clear. Such behaviour may have something to do with controlling the maturation process, or it could be that eels use some type of migration cue that is only found closer to the surface. More research is needed to yield answers to these questions.

The general hypothesis for eel migration is that silver eels leave Europe in autumn to spawn in the Sargasso Sea the following spring. Based on this belief and the distance they have to travel, the minimum migration speed can be calculated, but the speed of the PSAT-tagged eels in that study was less than the speed needed to complete the migration in that time. The reason for this can be either that the PSATs slow down the eels (as suggested elsewhere in this book), or that the eels take longer than hypothesized to complete their journey to the Sargasso Sea. European scientists, e.g. in the EU project EELIAD, have undertaken large-scale studies into eel migration, including alternative tagging methods. Experiments have been conducted with implanted tags, to preclude concerns that the eels may be slowed down by the drag caused by external attachment of tags. The principle in this work is that when the eel dies, after spawning or as a result of predation during migration, the tag would rise to the surface, drift shorewards and hopefully be recovered and returned to the address stated on the tag. Then, stored data such as temperature and depth will provide detailed

Fig. 3.17 *Top:* Inge Boëtius (86 in 2011) and Jan Boëtius (94) still have interest in the biology of eels and are happy to share their knowledge with other researchers. From left, guest editor Katsumi Tsukamoto, University of Tokyo, Jan Boëtius, Inge Boëtius, Jonna Tomkiewicz, DTU Aqua (photograph, guest editor Mari Kuroki, The University of Tokyo). *Bottom:* The well-known photograph "Eels in love", of a spawning-ready female European eel being courted by a smaller male eel in an aquarium in the experimental facilities of Inge and Jan Boëtius at the Danish Institute for Fisheries and Marine Research, now DTU Aqua (photograph, Inge Boëtius, Jan Boëtius and Paul Juhlin)

information on the general behaviour of the eels while they were migrating. Indeed, data from some of these smaller, implanted tags confirm both the speed and the unique migration behaviour of eels. This research is continuing and information is being gleaned on the consistency of the eel migration pattern, the route and the speed.

Research on Eel Reproduction in Captivity

Schmidt's discovery of the Sargasso Sea as the spawning area of the European eel not only evoked broad interest in eel spawning migration and reproduction, but also interest in understanding the complex hormonal mechanisms that control eel maturation and spawning. Danish researchers Inge and Jan Boëtius (Fig. 3.17 top) participated in a number of expeditions to the Sargasso Sea exploring the spawning area, but it was their experimental work on the reproduction of eels in captivity that caught the attention of scientists internationally.

Inspired by the pioneering work of the French scientist Maurice Fontaine, who succeeded in inducing maturation in male and female eels by hormone injection (Fontaine 1936; Fontaine et al. 1964), many attempts were made in Europe in the second half of the nineteenth century to reproduce eels in captivity. In Denmark, the maturation process and production of gametes was the focus of much of this research (Bruun et al. 1949; Boëtius et al. 1962; Boëtius and Boëtius 1967). Inge and Jan Boëtius studied hormonally induced maturation and the spawning behaviour of ripe male and female eels in aquaria. They were the first to describe eel mating behaviour, which they documented photographically, and their photograph "Eels in love" (Fig. 3.17 bottom) has been redrawn and published numerous times to illustrate the mating of eels. Their experimental work led to the first *in vitro* fertilization of European eel eggs in 1977 (Boëtius and Boëtius 1980), which then stimulated further work that resulted in Russian scientists successfully hatching European eel larvae for the first time in the 1980s.

The pioneering work on eel reproduction described above forms the basis for today's research into methods of breeding eels for aquaculture. Eel farms of today still base their production on wild-caught glass eels, but the ever-decreasing number of European glass eels arriving on our shores, similar to the situation for other species of eel, encourages eel aquaculture to move away from wild-caught eels towards a self-sustaining culture in which offspring are produced from captive broodstock. In the past decade, researchers in Japan have made significant advances in captive reproduction of eel, developing methods and technology to produce viable larvae and glass eels (Tanaka et al. 2001; Kagawa et al. 2005), and recently their eels have completed a full life cycle in captivity (Ijiri et al. 2011). The achievements thus far are promising in terms of future commercial production of glass eels under controlled conditions in culture, and the subject is covered in more detail elsewhere in this book.

The Japanese achievements around 2000 (Aida et al. 2003) led to renewed research on European eel reproduction, and the development of a self-sustaining culture of eels based on glass eel production from captive broodstock became a goal of the EU. The main bottlenecks to the production of viable European eel larvae are similar to those of the Japanese eel. They include poor egg quality through hormonally induced maturation, limited fertilization success, and embryonic and larval development failure. From 2005 to 2008, however, a Danish research team aiming to reproduce European eels in captivity improved egg quality and fertilization rates, and viable eel larvae were produced

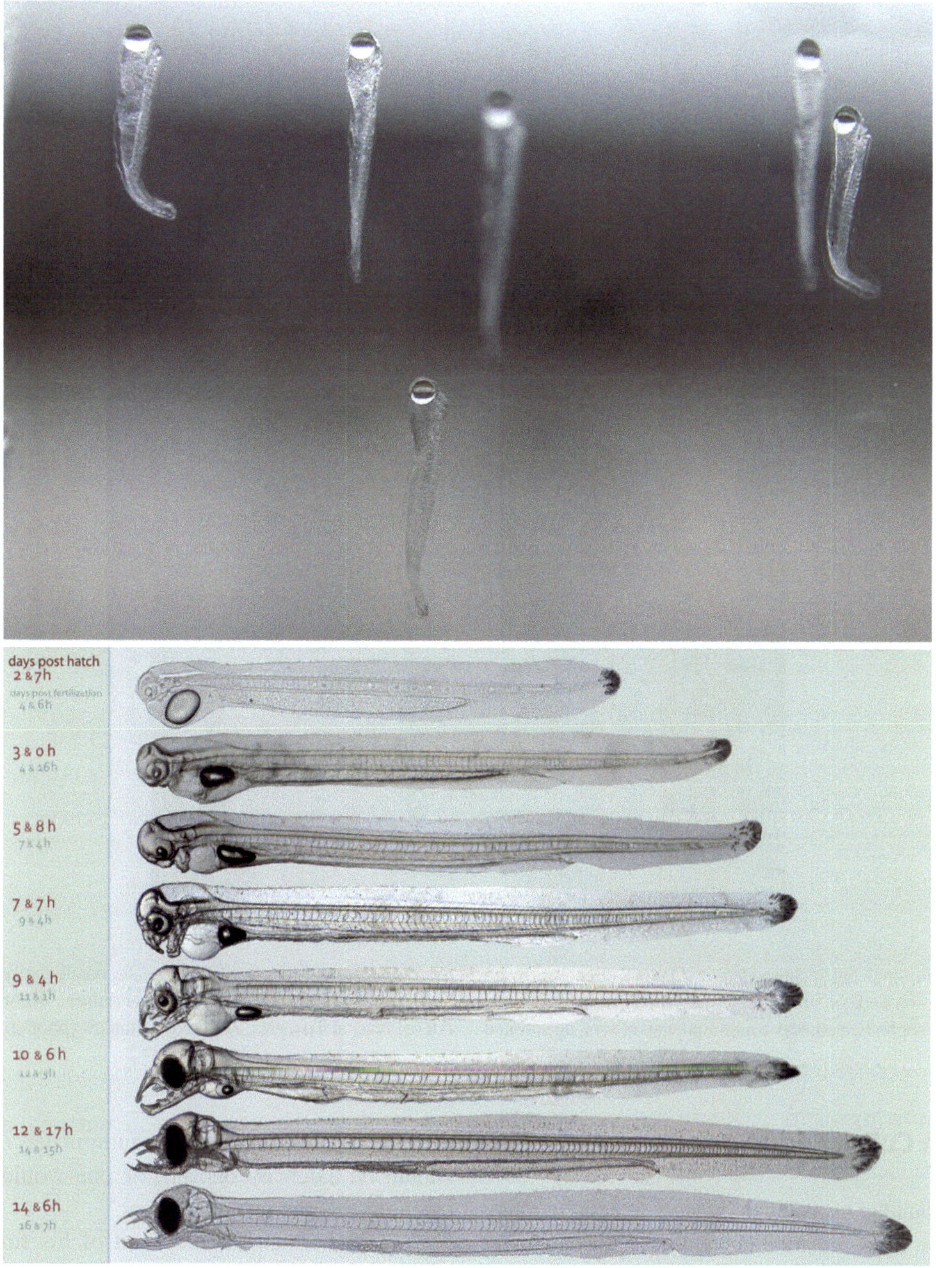

Fig. 3.18 *Top:* Newly hatched larvae of the European eel, the yolksac including a large oil droplet that helps a larva maintain neutral buoyancy. The bent newly hatched larvae straighten within a few hours (photograph, Sune Riis Sørensen, DTU Aqua). *Bottom:* Larval development in the European eel through the yolksac stage (0–12 days post-hatch) into the leptocephalus stage (14 day post-fertilization) in Danish experiments (photographs and graphics courtesy Jonna Tomkiewicz and Sune Riis Sørensen, DTU Aqua)

in large numbers (Fig. 3.18 top; Tomkiewicz and Jarlbæk 2008). The work led for the first time to European eel larvae completing their yolk-sac stage (pre-leptocephalus stage) in culture at an age of 12 days; during that stage, development depends exclusively on the nutrients deposited in the egg. More recently, in 2010, the same research team succeeded in culturing larvae in pilot feeding experiments, which then developed into the leptocephalus stage (Fig. 3.18 bottom; Tomkiewicz 2012). Currently, research on the reproduction of European eels is being fostered through the Danish-coordinated international research project PRO-EEL, which involves several leading research institutes and the aquaculture industry around Europe. A particular challenge now is to develop suitable feed for larvae and culture conditions appropriate for rearing leptocephali, a task that researchers in Japan continue to address in their efforts at developing a commercial production technology for Japanese glass eels.

Fig. 3.19 *Top:* Traditional eel dishes: (*left*) a Danish open sandwich ("smørrebrød") with smoked eel, omelette, tomato and chives on dark rye bread; (*right*) a more sophisticated, less-traditional way of serving smoked eel, with ice cream, oyster-filled tomato, parsley jelly, salmon roe and egg yolk (photographs, © Claes Bech-Poulsen, Kontrast ApS). *Bottom:* An eel feast at Tusenæs, Zealand, Denmark (photograph, Merete Ettrup)

Eels in Danish Culture and Society

Few fish are surrounded by as much mystery as eels. In many ways, eels do have some remarkable features. They can travel over land like snakes, their blood is poisonous, they migrate thousands of kilometres to spawn in the Sargasso Sea, and they have an uncanny, highly sensitive sense of smell. Likely because of some of these features, eels have been the subject of many sayings and myths over the years, but unlike the myths of some countries, whose myths suggest a more refined spiritual or religious leaning, Danish and other Scandinavian myths and sayings about eels often suggest a practical food-and-drink-related approach towards them. Some old sayings and myths heard in Denmark, along with some interesting suggestions for eel use in folk medicine are quoted below.

"If an eeler got an eel in the very first thrust, he might as well return home, since the yield would be poor. This seems lacking in logic, but fortunately the bad omen can be revoked by a drink; that is why one should always bring liquor when going eel fishing."

"The first caught eel of the autumn was not to be sold, but should be eaten by oneself, or one would have little luck catching eels the rest of the year."

"If a fisherman bit the head of his first eel, he would become 'King of eels', and eels would always seek him!"

"If you covered the rockers on a cradle with eel skin, the child could not be bewitched."

"Catholic priests who break the vow of celibacy will turn into eels after death."

"Drunkards can be treated with liquor in which eels had run themselves to death. The filtrated tincture should be drunk by the drunkard, who from then on would feel deep disgust for alcohol!"

"The sap of ash tree leaves mixed with eel fat and the juice of ant eggs can cure ear aches."

"Fresh eel can be eaten as a cure for constipation."

"Eel blood mixed with mint water can be drunk against colic."

"Eel skins can be placed on legs or joints to heal cramps."

Eels as Food

Eels are or rather were a traditional dish in Denmark. Because of their decline, however, few young people have grown up with eels as a common food, and it is mainly middle-aged and older people who like to eat eels. Eels are prepared and served in various ways. The traditional Danish ways to prepare eels were to salt large specimens of 1 kg or more, to smoke medium-sized eels of 350–1,000 g and to fry the smaller eels.

There are many different eel dishes in Denmark (two are shown in Fig. 3.19 top). Some of the more popular are fried eel with parsley sauce and potatoes, fried eel with stewed potatoes, smoked eel with eggs and wholemeal bread, eel soup, boiled eel with mustard sauce and potatoes, fried eel with stewed pear, fried eel with baked beetroot and apple, fried eel with herb cheese, and jellied eels. Typical herbs used in eel dishes are bay leaves, thyme, parsley, sage, tarragon, dill, chives and lots of pepper, but also garlic, curry powder and basil. Redcurrants and blueberries are a delicate topping sometimes used by top chefs.

Apart from individual dishes, eel feasts are traditional Scandinavian events, and they typically took place at the coast, where people used to support themselves by fishing for eels. They are feasts at which one can eat eel cooked in many different ways while consuming beer and schnapps (Fig. 3.19 bottom). The tradition of eel feasts began several hundred years ago, when the King owned all fishing rights and was willing to sell (for a royalty) those rights through local squires. Tradition had it that when the time was right and fishing for eels was at its most productive, the squires would arrive to collect the royalty, bringing potatoes and liquor, and the fishers would open their eel booths and offer the best of their fresh eels. One good story is that a true eel feast takes place when a group of men walks into an eel booth at dusk to eat and drink, then emerges from the booth in the early dawn with no recollection whatsoever of what had happened during the night! It is worth noting too that, because of local superstition or perhaps even morals, eel feasts used to be for men only, but nowadays both men and women participate in the festivities.

Final Comments

The eel was for eons one of the most important food fish in Denmark. However, its importance to the country and people is now dwindling simply as a consequence of declining abundance. Until the 1920s, the gear used to catch eels hardly changed and traditions did not alter, but since then, new gears, increasing fishing efficiency, the deterioration of eel habitats, barriers to both up- and downstream migration etc. have reduced the population and changed traditions. Focus in Denmark is now firmly on conservation of the stock, including finding means of sustainable exploitation and developing aquaculture.

Ongoing research into eel migration, reproduction, genetics and larval development is crucial to understanding the complexity of eel biology and hence managing these unique and fascinating fish. Equally important is to find means of preserving and improving habitats, and eliminating or at least lessening pollution and obstructions to up- and downstream migration. With successful reproduction of the European eel in captivity, the basis for self-sustained aquaculture will be laid, and in time too perhaps this can help improve the situation for the species in the wild.

As the European eel appears to consist of just one population, its future in Denmark is linked to the dynamics of the species Europe-wide and the ability of European people to work together to save it from extinction. Reinforcing this statement, ICES has stated that "The first priority is to get the message across to fishermen, managers, and politicians, that the most widespread and highest employing, single fish stock in Europe is dangerously close to collapse" (http://www.ices.dk/marineworld/eel.asp). The hope must therefore be that the EU Council Regulation of 2007 and national eel management plans will include measures sufficient to support the recovery of the European eel. Even if successful, however, any recovery will take a very long time.

Acknowledgements Maria Krüger-Johnsen and Lilian H. Andersen, DTU Aqua, collected information and photographs from archives, museums and researchers for the Eels and Humans exhibition in Tokyo that pre-dated production of the book and this chapter, and Sune Riis Sørensen, DTU Aqua, provided photos and videos for the exhibition and for this chapter. We also thank Søren Byskov, Fisheries and Maritime Museum, Esbjerg, Denmark, for providing information on historical fishing gear and tools, as well as artefacts for the exhibition, and Annette Jensen, Danish State Archives, for organizing and lending logbooks and letters pertinent to Johannes Schmidt's work. Photographs were provided by Thomas Højrup (Fig. 3.1), Rikke Johansen, the Viking Ship Museum, Roskilde, Denmark, Ingo Eichelberger and Robert Schabetsberger, Austria, guest editor Mari Kuroki, Claes Bech-Poulsen and Merete Ettrup. Last but not least, we thank Inge and Jan Boëtius for inspiring discussions and perspectives of the past.

References

Aarestrup K, Økland F, Hansen MM, Righton D, Gargan P, Castonguay M, Bernatchez L et al (2009) Oceanic spawning migration of the European eel (*Anguilla anguilla*). Science 325:1660

Aida K, Tsukamoto K, Yamauchi K (eds) (2003) Eel biology. Blackwell Publishing Asia, Melbourne, 497 pp

Als TD, Hansen MM, Maes GE, Castonguay M, Riemann L, Aarestrup K, Munk P et al (2011) All roads lead to home: panmixia of European eel in the Sargasso Sea. Mol Ecol 20:1333–1346

Anon (2008) Danish eel management plan. In accordance with Council Regulation (EC) No 1100/2007 of 18 September 2007 establishing measures for the recovery of the stock of European eel. © Ministry of Food, Agriculture and Fisheries, Denmark, December 2008

Boëtius I, Boëtius J (1967) Studies in the European eel, *Anguilla anguilla* (L.). Experimental induction of the male sexual cycle, its relation to temperature and other factors. Meddelelser fra Danmarks Fiskeri- og Havundersøgelser 4:339–405

Boëtius I, Boëtius J (1980) Experimental maturation of female silver eels, *Anguilla anguilla*. Estimates of fecundity and energy reserves for migration and spawning. Dana 1:1–28

Boëtius J, Boëtius I, Hemmingsen AM, Bruun AF, Møller-Christensen E (1962) Studies of ovarial growth induced by hormone injections in the European and American eel (*Anguilla anguilla* L. and *Anguilla rostrata* LeSueur). Meddelelser fra Danmarks Fiskeri- og Havundersøgelser 3:183–198

Bruun AF (1934) The life and work of Professor Johannes Schmidt. Estratto dalla Rivista di Biologia 16:3–22

Bruun AF, Hemmingsen AM, Møller-Christensen E (1949) Attempts to induce experimentally maturation of the gonads of the European eel, *Anguilla anguilla* L. Acta Endocrinol 2:212–226

Dannewitz J, Maes GE, Johansson L, Wickström H, Vockaert FAM, Järvi T (2005) Panmixia in the European eel: a matter of time. Proc R Soc Lond B Biol Sci 272:1129–1137

Drechsel CF (1890) Oversigt over vore Saltvandsfiskerier i Nordsøen og Farvandene indenfor Skagen, med Kort og Planer samt et Tillæg af Dr phil. C. G. J. Petersen [Overview of our salt water fisheries in the North Sea and the waters inside of Skagen, with maps, tables and a supplement by Dr phil. C. G. J.. Petersen]. A. E. Aamodt, Copenhagen, 151 pp

European Union (2007) Council Regulation (EC) No 1100/2007 of 18 September 2007 establishing measures for the recovery of the stock of European eel. Official J European Union L248(50):17–23

Fontaine M (1936) Sur la maturation complète des organes génitaux de l'anguille mâle et l'émission spontanée de ses produits sexuels [On the complete maturation of reproductive organs of male eels and the spontaneous emission of sexual products]. Comptes Rendus de l'Academie des Sciences 202:1312–1314

Fontaine M, Bertrand E, Lopez E, Callamand O (1964) Sur la maturation des organes genitaux de l'anguille femelle (*Anguilla anguilla* L.) et remission spontanee des oeufs en aquarium [Gonadal maturation of female eel (*Anguilla anguilla* L.) and spontaneous emission of eggs in aquarium]. Rendus de l'Academie des Sciences 259:2907–2910

Freyhof J, Kottelat M (2010) *Anguilla anguilla*. In: IUCN red list of threatened species, Version 2012.1. www.iucnredlist.org Accessed 5 August 2012

Grassi B (1896) The reproduction and metamorphosis of the common eel (*Anguilla vulgaris*). Proc R Soc London 60:260–277

ICES (2009a) Report of the 2009 session of the Joint EIFAC/ICES working group on eels. EIFAC Occasional Paper, No. 45. ICES Document CM 2009/ACOM: 15, Göteborg, Sweden, 7–12 September 2009, 540 pp

ICES (2009b) Report of the ICES advisory committee, 2009. ICES Advice, 2009, Book 9, 113 pp

Ijiri S, Tsukamoto K, Chow S, Kurogi H, Adachi S, Tanaka H (2011) Controlled reproduction in the Japanese eel (*Anguilla japonica*), past and present. Aquac Europe 36(2):13–17

Kagawa H, Tanaka H, Ohta H, Unuma T, Nomura K (2005) The first success of glass eel production in the world: basic biology on fish reproduction advances new applied technology in aquaculture. Fish Physiol Biochem 31:193–199

Kirkegaard E (2010) European eel and aquaculture. DTU Aqua Report 229, National Institute of Aquatic Resources, Technical University of Denmark, 19 pp

Munk P, Hansen MM, Maes GE, Nielsen TG, Castonguay M, Riemann L, Sparholt H et al (2010) Oceanic fronts in the Sargasso Sea control the early life and drift of Atlantic eels. Proc R Soc Lond B Biol Sci 277:3593–3599

Reagan TC (1933) Johannes Schmidt (1877–1933). J Cons Int Explor Mer 8:145–152

Schmidt J (1922) The breeding places of the eel. Philos Trans R Soc B 211:179–208

Sparrevohn CR, Storr-Paulsen M, Nielsen J (2011) Eel, seatrout and cod catches in Danish recreational fishing. Survey design and 2010 catches in the Danish waters. DTU Aqua Report, 240-2011, National Institute of Aquatic Resources, Technical University of Denmark, 22 pp

Tanaka H, Kagawa H, Ohta H (2001) Production of leptocephali of Japanese eel (*Anguilla japonica*) in captivity. Aquaculture 201: 55–60

Tomkiewicz J (ed) (2012) Reproduction of European eel in aquaculture (REEL). Consolidation and new production methods. DTU Aqua Report 249-2012, National Institute of Aquatic Resources, Technical University of Denmark, 48 pp

Tomkiewicz J, Jarlbæk H (2008) Kunstig reproduktion af ål: ROE II og IIB [Artificial Reproduction of Eel: ROE II and IIB]. DTU Aqua Report 180-08, National Institute of Aquatic Resources, Technical University of Denmark, 79 pp

Wind P, Pihl S (eds) (2004) The Danish red list, National Environmental Research Institute, Aarhus University. http://redlist.dmu.dk Accessed April 2010

Winge Ø, Tåning ÅV (eds) (1947) Naturforskeren Johannes Schmidt, Skildret af Venner og Medarbejdere [The Naturalist Johannes Schmidt: his life and expeditions: portrayed by friends and employees]. Gyldendal, Copenhagen, 187 pp

Wirth T, Bernatchez L (2001) Genetic evidence against panmixia in the European eel. Nature 409:1037–1040

European Eels: Dutch Fisheries, Culture and Eel Migration

4

Guido E.E.J.M. van den Thillart

The Netherlands is largely wetlands, so the country is classic habitat for the European eel (*Anguilla anguilla*), which was until relatively recently a common fish of the marine/brackish coastal area, found in rivers and wetlands. At high tide, half the country would be flooded were it not for the dikes and other water controls. Notably, many dikes, dams and sluices have been improved in the Netherlands during the past 100 years, to such an extent that migration of any fish became almost impossible. In 2008 the Netherlands had 4,671 pumping stations, 8,488 dams and 2,278 sluices, which given their advanced engineering, resulted in an almost complete barrier to fish migration (Fig. 4.1; Kroes et al. 2008).

A crucial part of the income of Dutch freshwater fishers was and still is derived from fishing for eels. Unsurprisingly, therefore, attempts were already being made 50 years ago to restock waters with glass eels to maintain a stable population in the Netherlands, at least until the price of glass eels burgeoned from the 1980s. Thereafter, the eel population declined rapidly all over Europe (Fig. 4.2; ICES 2009). The numbers of glass eels arriving at the coasts dropped dramatically throughout Europe, by about 90 % within 10 years, and since 2000, glass-eel arrivals have numbered <2 % of the number of arrivals during the 1960s. The decline in eel numbers started with a moderate decline in yellow eels some 5–10 years before a decline in glass eels was recorded, indicating that the downturn in the eel population in Europe was almost certainly not attributable to oceanographic changes, but was based on changes in coastal and wetland habitat. In this respect, it is remarkable that the same pattern was observed for American, Japanese and European eels over the same period, although this does not mean that the cause was necessarily the same for all three species. However, because the habitats of the three eel species are all within well-populated, urbanized areas, it is likely that a complex of factors related to urbanization impaired all three.

In the past, many fishing villages in the Netherlands, particularly those adjacent to IJssel Lake, such as Volendam, Spakenburg, Urk, Harlingen and Lemmer, relied heavily on eel fishing, but as the eel population declined in the 1980s, fishers either moved onto other fish species or stopped fishing altogether. Still, however, there is evidence in all those villages of a past and present interest in eels, for instance in the maintenance of old fishing boats (Zuiderzee Botters) mainly now for recreation (Fig. 4.3). In Volendam, there is also an old building called "De Visafslag," which was still in use some years ago as a place to auction eels caught in Lake IJssel (Fig. 4.4). Lake IJssel was formerly known as "Zuiderzee" (IJssel Sea), an inland water body for which plans to close its link with the sea already existed in 1819. Real efforts to close it, however, started in 1912, with the actual engineering only commencing in 1927 with the construction of a dike 30 km long and 90 m wide. The construction work was completed in 1932, and the lake is now the largest in the Netherlands, with a surface area of 1,100 km^2.

Eels in the Netherlands were traditionally processed in several ways, although the preferred one was smoking the catch from Lake IJssel; that catch consisted largely of small male eels (30–40 cm long and 80–150 g) which were available year-round. Now, however, the much smaller catch from the lake is exclusively female, because of the negligible influx of glass eels today, showing that the sex ratio is primarily determined by population density. The smaller male eels processed these days are derived exclusively from farms. Most of the Dutch eel catch is now processed by a few large companies, still mainly in the form of smoked product (Fig. 4.5). Eel smoking is, however, a national pastime, many people know how to do it, and there are still contests for the best smoked eels. Indeed, even with the species IUCN red-listed as endangered, sports fishers rarely return their catch of eels to the water, in contrast to what they do with other fish species. Instead, they take the fish home and fry or smoke them.

G.E.E.J.M. van den Thillart (✉)
Institute Biology Sylvius Laboratory, Leiden University, Postbus 9505, Sylviusweg 72, k4614, 2300RA Leiden, The Netherlands
e-mail: g.van.den.thillart@biology.leidenuniv.nl

K. Tsukamoto and M. Kuroki (eds.), *Eels and Humans*, Humanity and the Sea,
DOI 10.1007/978-4-431-54529-3_4, © Springer Japan 2014

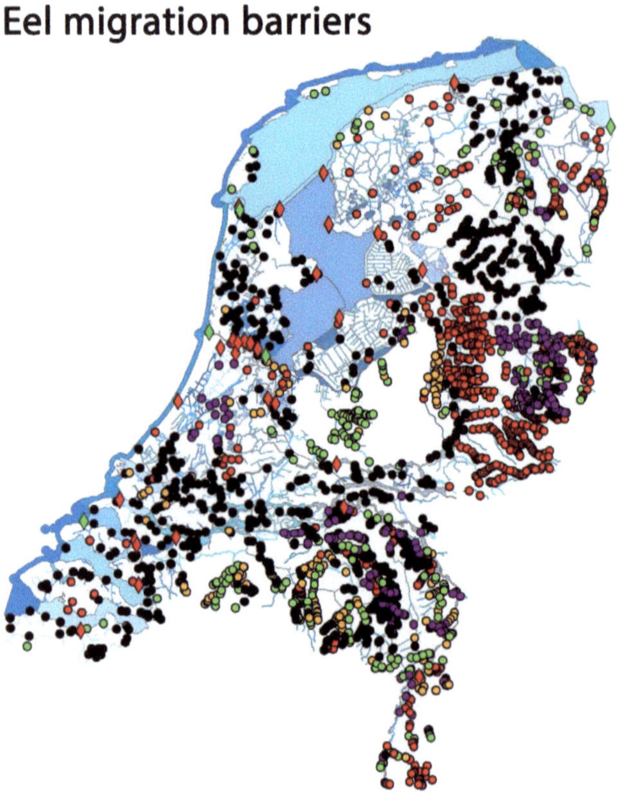

Fig. 4.1 The major barriers to eel migration in The Netherlands. The waterways are virtually closed to all fish migration by 4,671 pumping stations, 8,488 dams and 2,278 sluices (in 2008; modified from Ministry of Agriculture and Nature and Food Quality 2009). *Diamonds* depict barriers with (*green*) or without (*red*) a fish passage; *dots* depict critical sites with a passage in existence in 2008 (*green*), 2010 (*orange*), predicted by 2015 (*red*), predicted to be provided but date not yet known (*purple*), or with no passage currently planned (*black*)

Fig. 4.3 (**a**) Traditional fishing boats ("Zuiderzee Botters") still sail on IJssel Lake, but just for recreation. The boats are built for shallow water; they have side boards that are lowered when the winds are from the side to act as a keel. (**b**) Many of them are berthed in Spakenburg, an old Dutch eel-fishing village

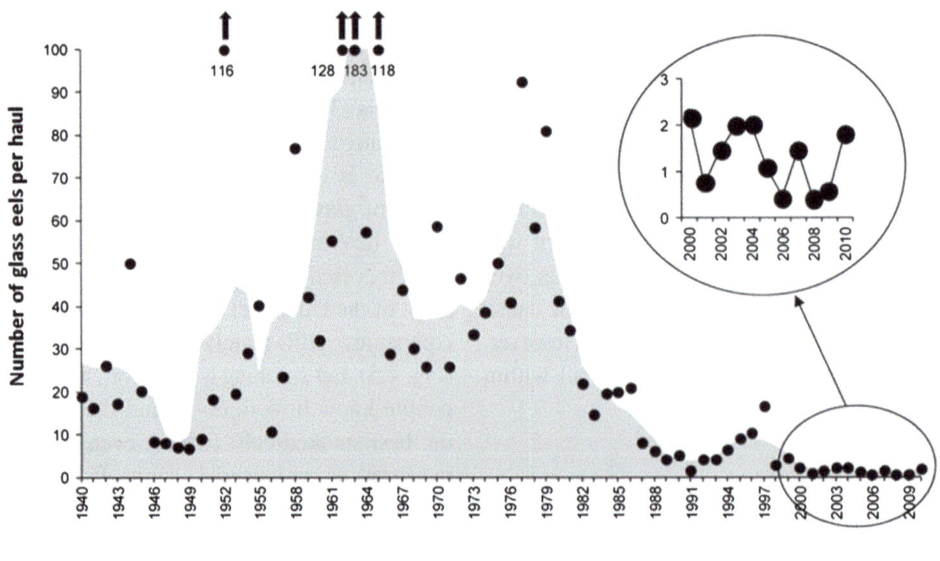

Fig. 4.2 Decline of the eel population in the Netherlands reflected in the results of glass-eel surveys, 1940–2010 (modified from de Graaf and Bierman 2010). The dots are the annual index, and the grey area the running 5-year means

Fig. 4.4 A "visafslag" in Volendam, a Dutch village well known for its eel fishing tradition. The old building was used exclusively to auction the fish brought in by the many IJssel Lake fishers. It was still in use up to a few years ago

Fig. 4.5 Eels in the Netherlands are mainly sold in smoked form, the favoured product. European eels have a high fat content, 20–30 %, which make them ideal for smoking

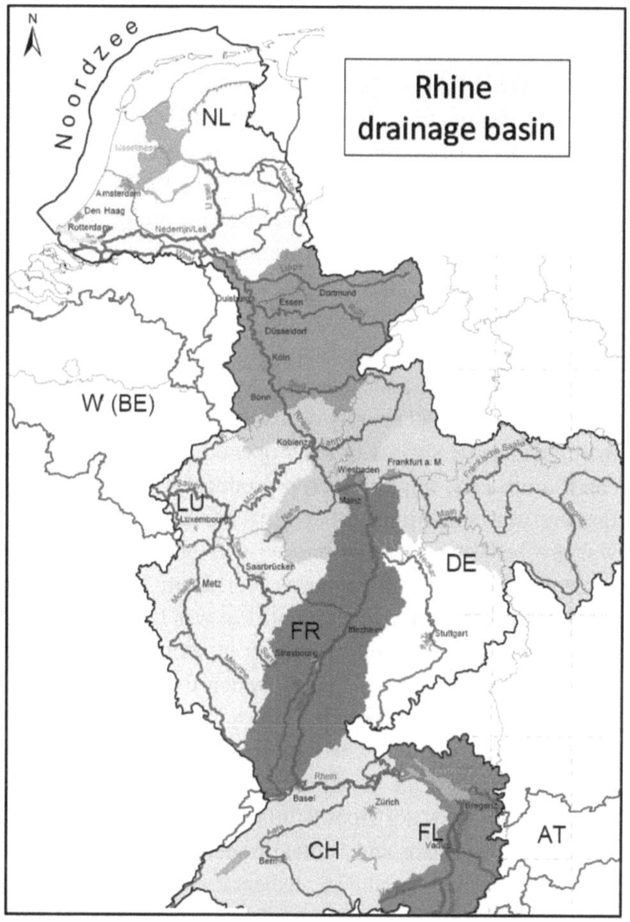

Fig. 4.6 The drainage basin of the Rhine, the largest river in Europe, which runs mainly through Switzerland (CH), Germany (DE), France (FR) and The Netherlands (NL), the last made up mainly of the river's delta. Smaller rivers such as the Scheldt, Maas and Elbe also contribute to the delta and wetlands (modified from Ministry of Agriculture and Nature and Food Quality 2009). BE, Belgium; LU, Luxembourg; AT, Austria; IT, Italy

Although fykenets alone are deployed by commercial fishers, longlines and "peuren" (bobbing) are used occasionally by individual fishers in the wetlands. Longlines with a number of hooks are tied to a small stick placed in the grass in a ditch, where they are virtually impossible to detect. The tradition of "peuren" is a long-established means of fishing for eels, and it is still carried out by some as a sport; in Leiden there is an annual "peurbakken tocht," a festival at which everyone shows up with their (well-decorated) boat. The bobbing technique relies on the typical bite response of eels, which do not let go of bait easily. For bait, rain worms are put on a line and bundled into a ball, connected to a weight on the line, and moved up and down with a stick just above the bottom. When an eel bites the bundle, it holds onto it tightly provided it is pulled carefully up in the water and into the boat.

History and Fishing Rights

The Netherlands span the delta of several rivers flowing into the sea along the west coast of Europe, the Rhine being the most important (Fig. 4.6). However, the wetlands were already well drained and the water level controlled by dams and sluices many hundreds of years ago, seawater and freshwater exchange taking place through sluices, and migratory fish such as eels readily passing the simple barriers, making sluice-fishing profitable. In fact, as long as 1,000 years ago, most eel fishing was associated with sluices. Indeed, in the Netherlands, eels have been historically far more important in inland waterways than salmon, and the fishing rights for eels were already well regulated and the exclusive right of feudal lords in the Middle Ages. The species was an important source of income to most of the counts of Holland (which is in the western part of the Netherlands), delivering as much as 20–40 % of their annual stipend (van Dam 1997).

The water level in the wetlands was generally controlled by monasteries that introduced drainage canals and sluices in order to encourage the people to remain living locally. However, the sluices then were not as effectively engineered as they are today, often leaking badly, allowing glass eels to enter and yellow and silver eels to exit with relative ease. As fishing rights had priority over water control, fishers often manipulated the sluices or even opened them to be able to catch more eels, impairing crop production and infuriating farmers and sometimes other citizens. Such almost open warfare between affected parties was at its worst in the fifteenth and sixteenth centuries (van Dam 1997). During the economic upturn then, a lot of peat was being dug in the wetlands to fuel the cities, resulting in serious land erosion and the formation of large lakes, which then also increased the habitat available for eels. The most successful sluice-fishing was carried out where the water drained into a river and/or the sea, and the most productive drainage point was at Haarlem, where water from Rijnland (the area between the cities of Amsterdam and Delft) drained into the Zuiderzee (now IJssel Lake). At sluices, fishers used all forms of fishing gear suitable for catching eels: fykenets, wicker pots, wicker fences and even trawls. Later, after 1500, polenet fishing was introduced, but that required investment beyond the means of many. Much of the trade in eels taking place at the end of the Middle Ages was in smoked and live product, and large quantities of live eels were transported in specially constructed barges to London, where Dutch traders sold their eels at the market (van Dam 1997).

Nowadays with tightly fitting sluice doors, high-frequency pumps and water flows irregularly directed into the wetlands, virtually no eels can enter or leave Dutch waterways and polders. In the past, the pumps were driven by windmills (Fig. 4.7a), with loosely fitting paddles transporting water over a small height difference, usually <1 m, which is no real barrier to an eel. Most lakes were shallow and often small, and they were the first to be drained using simple windmills. Having initially destroyed much of their land mining peat, the new technology using windmills allowed local populations to reclaim their lost land, and farmers around lakes started to form cooperatives. These organizations, referred to as "waterschappen" in Dutch, often constructed dikes and canals around their lake, then built one or two windmills to drain the main lake, referring to the reclaimed land as a "polder." Once the land had been reclaimed and a polder formed, the local cooperative was responsible for its maintenance, and waterschappen had already by the fifteenth century a status similar in officialdom to that of a city council. Such organizations still exist today, with the same responsibilities, each waterschap collecting taxes from its local inhabitants and using the income so derived exclusively for waterworks and water control. Waterschappen were the first experience of a democratic system in the Netherlands, the people living near the polder voting their representatives onto the council. Nowadays, though, some waterschappen have merged to increase their efficiency and supporting their ability to construct and maintain the large, complex engineering works required for modern society.

Fig. 4.7 (**a**) The Red Mill in Westlagelandspolder, now called the "Rode Polder," was built in 1632. It is close to Leiden (in the village of Oud Ade) in the area where the first lakes were drained using windmills. The upper part of the mill, the house, can be turned and positioned with beams on the outside to be aligned with the wind direction. (**b**) The "Cruquius" was the first steam-powered pumping station in the Netherlands and was one of the three steam-powered stations built to drain the last and largest lake in the Netherlands, the Haarlemmer Meer, which was for many years a threat to nearby cities

The largest lake in the Netherlands was Haarlemmer Lake, and relative to other lakes it was deep and large, at 18,000 ha. In fact, it was too large to be drained with windmills and was a threat to cities in the area, growing larger every year as a result of wave action during storms. Sometimes during those storms, the water destroyed the dikes, and almost every year a wide area was newly flooded. It was not until the nineteenth century, however, that an engineering solution became available with the design and construction of powerful steam-driven pumping stations at three locations around the lake. It then took 4 years to drain the lake, between 1848 and 1852. One of the steam-powered pumping stations, Cruquius, has been saved and is now a museum (Fig. 4.7b). The reclaimed land from Haarlemmer Lake was initially used for agriculture, but it now carries Amsterdam (Schiphol) airport, and the balance of the land is used to grow flowers and to accommodate high-tech industry and expanding towns.

Eel Fisheries, Management and Aquaculture

The European eel is in decline throughout its range and all current fisheries are considered outside sustainable limits. As mentioned elsewhere in this book, factors involved in the decline include overexploitation and other anthropogenic impacts (habitat loss, barriers to migration, pollution) plus natural impacts (predation, increasing parasite loads, climate change). Human-induced mortality is high on both juvenile (glass) and older (yellow and silver) eels, and eel recruitment throughout Europe is at a historical low and continuing to decline, with no obvious signs of recovery. Current levels of mortality induced by anthropogenic activity in the Netherlands and elsewhere are clearly unsustainable, so there is an urgent need for them to be reduced as low as possible until some recovery is recorded. Glass-eel recruitment data demonstrate a clear decline since the early 1980s, levels having dropped to between 1 and 9 % of the levels in the 1970s.

The paragraph above is based largely on a statement in the EIFAC/ICES working group on eels report of 2003 (ICES 2009), and the gravity of the situation regarding eels was soon noted by the European Commission, which in 2009 obliged Member States to implement an eel management plan for each water basin, to protect the eel population in Europe. The eel population throughout the continent decreased fast in the 1980s (Fig. 4.2), but because the real cause of the decline was not evident, it took a long time for mitigating action to be taken. Clearly, however, many factors played a role in the virtual collapse of the eel population, and it is probably the number of factors and indecision on which were the most important that made it difficult to decide on the action necessary to halt the decline and to attempt to stimulate recovery. One contributing factor has certainly been predation by cormorants, because such predation is known to have accounted for 30–50 % of the whole take of European eels during 1993/1994. Another contributing factor was the arrival in the early 1980s, likely with a shipment of Japanese eels (*Anguilla japonica*), of the eel swimbladder parasite *Anguillicola crassus*. That infestation spread fast in western Europe, probably because the European eel had no natural resistance to it. Indeed both the levels of infestation and the sizes of the parasite in European eels outstrip the levels recorded in Japanese eels. The parasite is clearly more destructive to the European eel than it is to the Japanese eel, because it not only impairs swimbladder function but also feeds on the blood of the host eel (Fig. 4.8). After repeated infections, the eel swimbladder becomes thick and ultimately constricts completely, such that it cannot be inflated by the gas gland, which has an obviously devastating effect on silver eels migrating to their spawning grounds across the Atlantic and conducting regular diurnal migrations. Without a functional swimbladder, the energy costs of swimming are very high (Palstra and van den Thillart 2010).

As stated above, the European Commission obliged its Member States to produce eel management plans with the aim of reducing mortality on the species and preserving the stock. With an estimated total production of eels in the Netherlands during 2009 of some 6,120 t (commercial fishing 920 t, sportfishing 200 t, aquaculture 5,000 t), the national plan implemented in that year is articulated in Ministry of Agriculture and Nature and Food Quality (2009). In summary, it stated that:

- there would be no fishing during the migratory season (September–November);
- 30 important barriers to migration would be removed;
- 600 fish passages would be created before 2015;
- a 35 % reduction in mortality at power plants would be sought;
- a restocking programme with glass eels would be established.

The objective of the plan was to ensure that 400–5,200 t of silver eels would be migrating annually into the North Sea from the Netherlands by the year 2090. The target is an escapement rate of silver eels of at least 40 % of the pristine situation, taken as that in 1970, and until that level was achieved, all fishing for eels in Europe is banned. In the Netherlands in 2010, all fishing for eels during their migration months (September, October and November) was stopped. To address the restocking target for glass eels, a group of eel fishers, eel farmers and members of the eel processing industry (in an organization referred to as "Future for Eel," and subsequently as DUPAN) tried first to improve negative public sentiment about the threatened eel population because they were worried about both the reduced yield from the stock and the strict measures taken by Government. However, to make matters worse, Dutch supermarkets in 2010, influenced by NGOs such as WWF, Stichting Noordzee and Greenpeace, banned all products based on the European eel, and as a direct consequence of that decision, FFE/DUPAN started releasing silver eels into the North Sea (Fig. 4.9), arguing that attempting to restock with glass eels could not provide immediate help to the decimated eel population, because it took >12 years before surviving glass eels migrated as silver eels back to their spawning grounds. In the opinion of that group, therefore, releasing large female silver eels would be a more effective boost to the

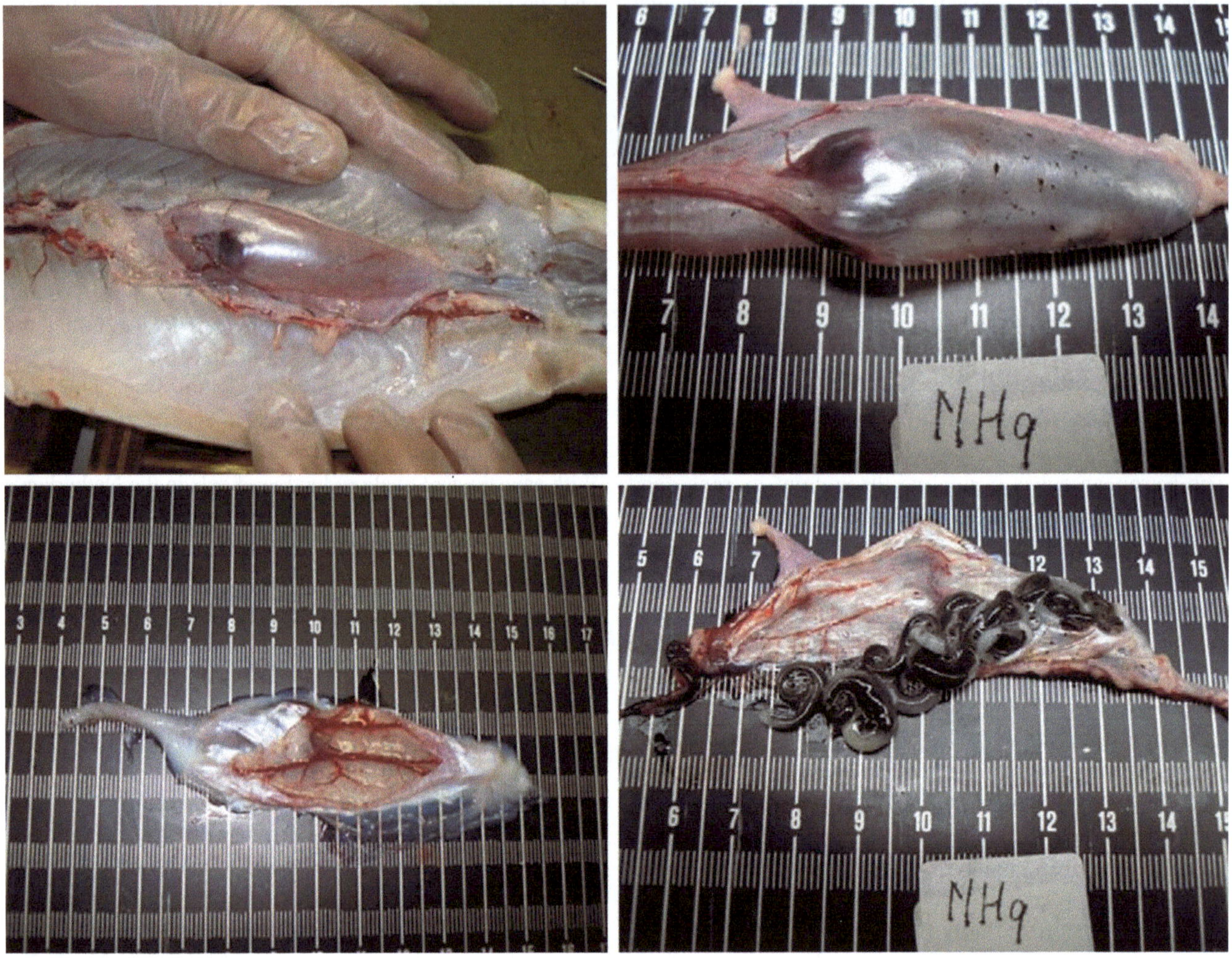

Fig. 4.8 The swimbladder parasite *Anguillicola crassus*, a nematode that uses copepods and other fish as intermediate hosts. The larvae travel from the gut of the eel to the swimbladder, where they mature. Eggs leave the swimbladder for the gut via the pneumoduct. After repeat infections, the swimbladder wall thickens and cannot be inflated by the gas glands

Fig. 4.9 Large-scale release of silver eels into the North Sea by the "Future for Eel" foundation (FFE)

stock and might result in increased glass eel influx within as little as 2 years. The jury is still open on whether this effort will improve the situation!

Sports fishers, unlike commercial operators, rarely fish in open waters. Instead, they operate mainly from small boats or along the banks of canals, generally with rods. Commercial fishers use different techniques to catch eels, operating in rivers and open water mainly with fykenets, often with several connected to each other over a distance of several hundred metres and fixed with anchors (Fig. 4.10). Such gear is generally set along the side of the river, to preclude it being damaged by passing cargo ships. The modern boats used to catch eels differ greatly (Fig. 4.11) from traditional ones, but they still look impressive when operating in open water under a clear sky.

It was some 30 years ago that interest in eel aquaculture in the Netherlands burgeoned, and many pig farmers then

Fig. 4.10 Commercial eel fishers, fykenets and fishing boats. Fykenets are anchored and connected to each other along the river bank over a distance sometimes of >1 km

Fig. 4.11 Fishing boats in the Netherlands

Fig. 4.12 A modern eel farm with automatic feeders and water recirculation system. Eels can be held successfully at high density, but the major issues are to keep water flow rapid and water quality good (photo courtesy J. van Doren, reproduced with permission)

started their own eel production plant, using recirculation techniques and automatic feeding systems that allowed the plant to be run virtually automatically in confined spaces in small buildings (Fig. 4.12). Water recirculation has the advantages of limited heat loss, constant water condition and the ability to culture fish at high density, meaning of course less need for high-cost space and labour. Although profits have dropped because of the rising purchase price of glass eels, feed, water and electricity, these eel farms still produce more eels than the fishery. Eels at these farms are harvested after about 1–3 years when they have already grown to marketable size and when they tend to stop feeding and are about to adopt the silver phase. The farmers do not want their product to attain the silver phase, because it is more efficient economically for them to harvest the eels at the end of their main growing phase, so they then sell their product to processors, who mainly fillet and smoke it before any silvering takes place. Some eels are also sold whole smoked, the way the product was traditionally produced in the Netherlands (Fig. 4.13), but by far the largest part of eel production today is filleted smoked eel sold in sealed packaging. In the Netherlands, the smoking of eels is considered an art, so many smaller companies sell their product as "produced in the old-fashioned way." However, bigger processors have the capacity to smoke large numbers of eels, so they import much of their product from elsewhere in Europe (Norway–Greece), using special trucks and holding the imported eels temporarily in an adjacent canal (Fig. 4.14). These bigger processors have a broad international network for import and export. Much of the product sold on is smoked, often in the old fashioned way, i.e. over smouldering wood chips that guarantee the exclusive flavour of Dutch smoked eel (Fig. 4.15).

Fig. 4.13 Smoked eels are sold whole and filleted, the latter becoming more popular as people become increasingly unwilling for their hands to be greasy after eating eels. Photos courtesy J. van Doren (*left*) and A. Koelewijn (*right*), reproduced with permission

Eel Migration–the Energy Requirements of Swimming

European eels have to cross the Atlantic Ocean twice: on their way to Europe they mainly drift passively in the Gulf Stream as leptocephalus larvae, arriving along the coast and entering the lower reaches of estuaries as glass eels and elvers, but when the survivors return as adults to the Sargasso Sea to spawn, they have to swim the ~6,000 km. Although the energy expenditure required for the massive return migration is enormous, the life history of the species does provide it with the ecological advantage of being able to be distributed over a wide area of coastal Europe, from Norway to Morocco and even into the Mediterranean Sea, as far afield as Greece and Egypt. This pattern of an outward passive drift of larvae with currents and an inward active spawning migration as adults is an innate characteristic of the species, because all 19 species of *Anguilla* (Minegishi et al. 2005) show it. However, only a few species migrate such great distances into the open ocean as the European eel *Anguilla anguilla*: these are *A. rostrata*, *A. japonica*, *A. australis*, *A. dieffenbachii* and *A. marmorata*. The shortest distance covered among these six species is that of *A. australis* and the longest that of *A. anguilla*, respectively ~2,000 and ~6,000 km.

Migrating eels do not feed, so they rely for their energy on fat stores which, when they leave freshwater, constitute as much as 30 % of body weight (Tesch 2003; van Ginneken et al. 2005; van den Thillart et al. 2007). European silver eels need to swim across the Atlantic Ocean within 6 months, because that is the difference in time between the date they leave the coast and the date the first larvae are observed in the Sargasso Sea. Therefore, minimum swimming speed would be 6,000 km in 6 months, or 0.4 m s^{-1}. To date, little is known about swimming speeds and the endurance of eels, and these parameters, plus the rate of oxygen consumption, are required for scientists to be able to answer questions about the energy requirements for eels to swim such long distances. Long-term swimming experiments require specialized equipment that needs to be able to run continuously for several months, and it is necessary too to be able to determine energy consumption at different swimming speeds, because there obviously needs to be sufficient energy left for the eels to be able to spawn when they eventually arrive at the position where they breed. From energy consumption values of other fish species, the energy stores of an eel (measured at 300 g of fat per kg of eel) would be depleted before 6,000 km of migrating without feeding. Hence, eels need to have an efficient means of swimming.

To study the swimming energetics of eels, 22 swim tunnels 2 m long were built and tested with female eels ~80 cm long (van den Thillart et al. 2004). The tunnels were tested for

Fig. 4.14 Live eels are imported to the Netherlands by truck and either kept in nets in a nearby canal or used immediately. A stunning technique, based on short electrical pulses, is now used to kill them humanely, and thereafter they are treated with salt to remove the slime before their gills and intestines are removed

Fig. 4.15 After cleaning, eels to be smoked are put on spikes and placed in large racks above smouldering woodchips that impart the traditional flavour. During the smoking process, much of the fat melts

homogeneity of water flow, calibrated with a laser–Doppler technique (van den Thillart et al. 2004). Extensive endurance swimming trials were carried out at Leiden University (van Ginneken and van den Thillart 2000; van Ginneken et al. 2005; Palstra et al. 2008; van den Thillart et al. 2009). The tunnels were of the Blazka type, consisting of two concentric tubes, with the inner tube and the outer ring having the same surface area, resulting in equal flow rates in the two compartments. A propeller pushes water into the outer ring and further into a bundle of flow streamers, to reduce the size of the vortices and generating a semi-lamellar flow (Fig. 4.16). Eels are obviously excellent swimmers, because placing them in flowing water is usually sufficient for most of them to start swimming actively. However, it is important in such experiments to preclude the fish from becoming stressed when being induced to swim; gradually altering the flow rate up and down gently during the first hour after experimental start-up is generally sufficient to stimulate them to swim. Once swimming, eels can swim for long periods at speeds of 0.4–0.8 body lengths (BL) s^{-1}. An endurance swimming trial has been performed with adult female eels swimming for 160 days at 0.5 BL s^{-1}, which would equate to a horizontal distance covered of 5,500 km. The energy consumption by such eels swimming continuously seems to be low, about 5 times less than that of salmonids (van Ginneken et al. 2005; van den Thillart et al. 2009), likely a selective force during evolution of the species. Despite male European eels being much smaller than females, 40 cm vs. 80 cm, they swim as well as females, and their energy costs of swimming appear to be even lower. Moreover, tested swimming in groups (Fig. 4.16c), the energy cost of swimming by males is even less, some 20–50 % less than when swimming alone (Burgerhout et al. 2013).

In experiments conducted in Leiden in swimming gutters, where it is possible to assess the swimming capacity of large groups, it has proven possible to study endurance swimming relative to sexual maturation. It is of note that silver eels leaving Europe in autumn are not yet mature, with a GSI (gonadosomatic index) of 1–2 %, and seldom >2 %. Therefore, the migrating eels must be maturing later, either during their migration across the Atlantic or perhaps when they arrive in the Sargasso Sea. One theory is that sustained swimming over a long period is a trigger that releases the inhibition against maturation exerted by certain brain centres. Results of experimentation to date do indicate that swimming is a trigger for maturation; males seemingly mature after swimming continuously for a few months and females show initial changes in the ovaries during the long swim. However, females clearly require an additional trigger to stimulate maturation. Knowledge of natural triggers is certainly important for future work on stimulating eels to spawn in captivity, because the repeated hormone injections upset the condition of the eel and are both expensive and time-consuming. Another way perhaps to improve eel reproduction under culture conditions would be to stimulate natural spawning behaviour

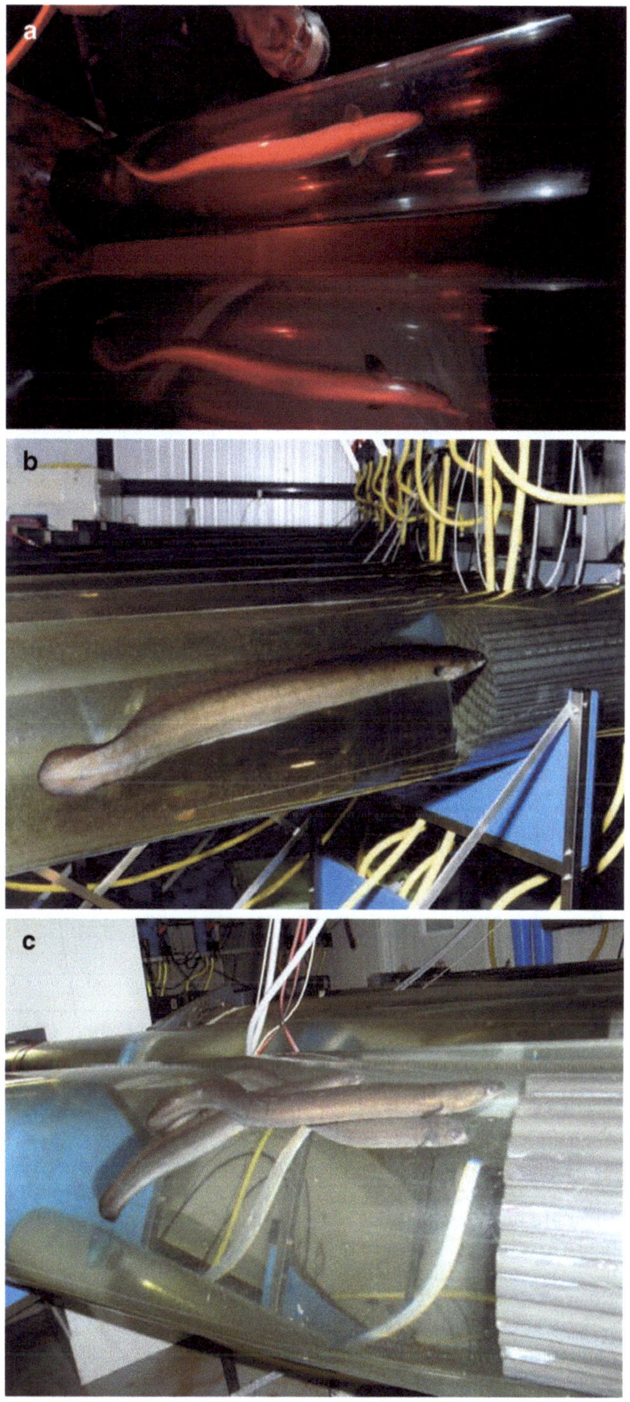

Fig. 4.16 Testing eel swimming stamina in a swimming tunnel consisting of two concentric perspex tubes 2 m long. (**a**) Eels in the tunnel, with red light used to limit stress–silver eels are virtually blind to red light (photo Doubilier, reproduced with permission). (**b**) Swimming tunnels are used to study long-distance migration. In the foreground is a tunnel with a 72-cm silver female eel. (**c**) Male eels, which are smaller than females, swimming in groups, which reduces energy expenditure (van den Thillart et al. 2004)

(Fig. 4.17); mature males and females induce the final phases of maturation through pheromone signals. Natural maturation would probably be more effective than manual

Fig. 4.17 The spawning behaviour of mature female and male silver eels. The female eels were treated with pituitary extract from carp and male eels with hCG (human Chorionic Gonadotropin) to induce final maturation (photo courtesy H. Berkhout, reproduced with permission)

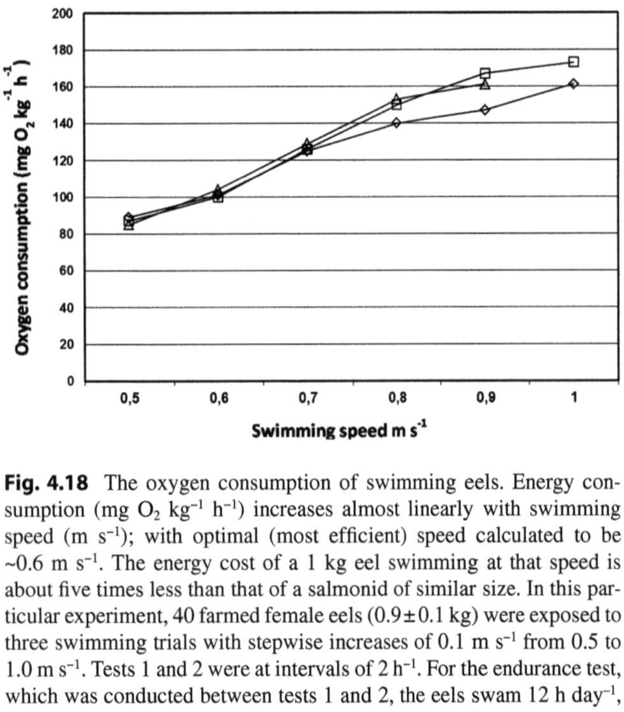

Fig. 4.18 The oxygen consumption of swimming eels. Energy consumption (mg O_2 kg^{-1} h^{-1}) increases almost linearly with swimming speed (m s^{-1}); with optimal (most efficient) speed calculated to be ~0.6 m s^{-1}. The energy cost of a 1 kg eel swimming at that speed is about five times less than that of a salmonid of similar size. In this particular experiment, 40 farmed female eels (0.9 ± 0.1 kg) were exposed to three swimming trials with stepwise increases of 0.1 m s^{-1} from 0.5 to 1.0 m s^{-1}. Tests 1 and 2 were at intervals of 2 h^{-1}. For the endurance test, which was conducted between tests 1 and 2, the eels swam 12 h day^{-1}, each consecutive day at a faster speed (modified after Palstra et al. 2008). Diamonds, test 1; triangles, endurance test; squares: test 2. Differences between the three tests were not significant

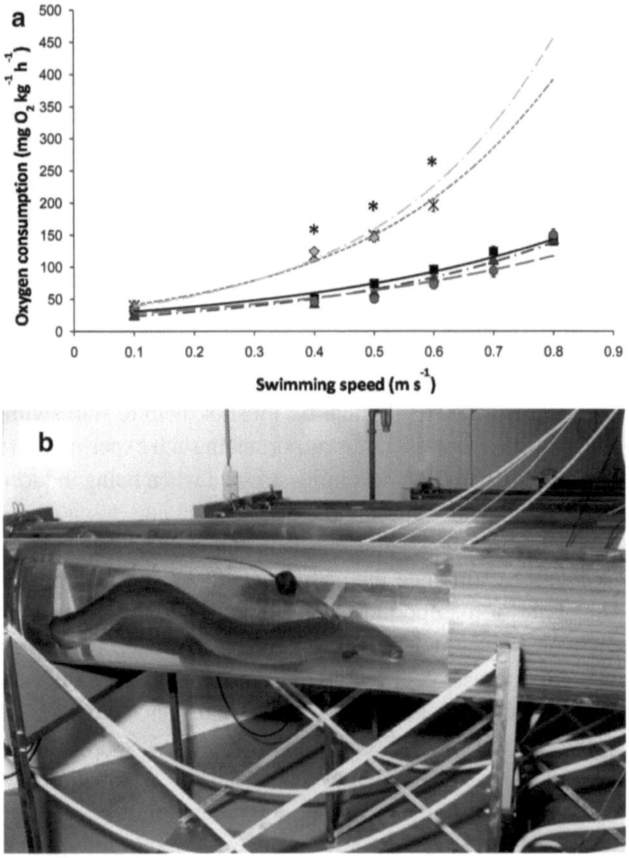

Fig. 4.19 An eel's swimming efficiency and endurance drop markedly when a tag is attached. (**a**) The rate of oxygen consumption (mg O_2 kg^{-1} h^{-1}) of European eels (~0.8 kg) swimming at 0.1, 0.4, 0.5, 0.6, 0.7 and 0.8 m s^{-1} with/without an attached pop-up satellite tag (PSAT), i.e. under five different conditions: no tag (*black squares*), after attachment of a base for the tag (*grey circles*), with tag attached (*grey diamonds*), with the same tag made neutrally buoyant (*crosses*), and after removal of the tag (*grey triangles*). The asterisks indicate significant differences ($p < 0.01$) between eels swimming with and without a tag. Eels carrying a tag could not swim faster than 0.6 m s^{-1}. The error bars (± s.e.) are indicated at each datapoint. Graph modified after Burgerhout et al. (2011). (**b**) A 1 kg European eel with a PSAT attached to it in the swimming tunnel

induction of maturation with injections of ovulation hormone such as DHP (17, 20-dihydroxy-4-pregnen-3-one).

To determine optimum swimming speed, experiments were carried out with eels swimming for extended periods at different swimming speeds (Palstra et al. 2008). Four groups of eels were tested in a swimming fitness protocol over a range of speeds of 0.5–1.0 m s^{-1}, corresponding to 0.6–1.2 BL s^{-1} (Fig. 4.18). In that study, eels were allowed to swim for 2 h at each speed from 0.5 to 1.0 m s^{-1} increasing in steps of 0.1 m s^{-1}. At each speed, oxygen consumption was measured continuously for 90 min and the maximum aerobic speed interpolated according to the method of Brett (1964). A group of 40 farmed eels was tested twice (tests 1 and 2) at 2-h intervals; each test took one day. In the week between tests 1 and 2 too, each eel was required to swim for 12 h day^{-1}, each day at a faster rate, i.e. 0.5, 0.6, 0.7, 0.8 and 0.9 m s^{-1}, to assess endurance. The rates of energy consumption were the same in the different tests, and it was concluded that, once eels start to swim, they do not alter their swimming mode, and more importantly, that the energy costs of swimming do not change. Eels are efficient swimmers, more so than many other species of fish, but when swimming they do not cope well with tags attached externally. Recently, PSATs (pop-up satellite tags), which have been used a few times to follow migrating eels (Jellyman and Tsukamoto 2002; Fig. 4.19), were tested in

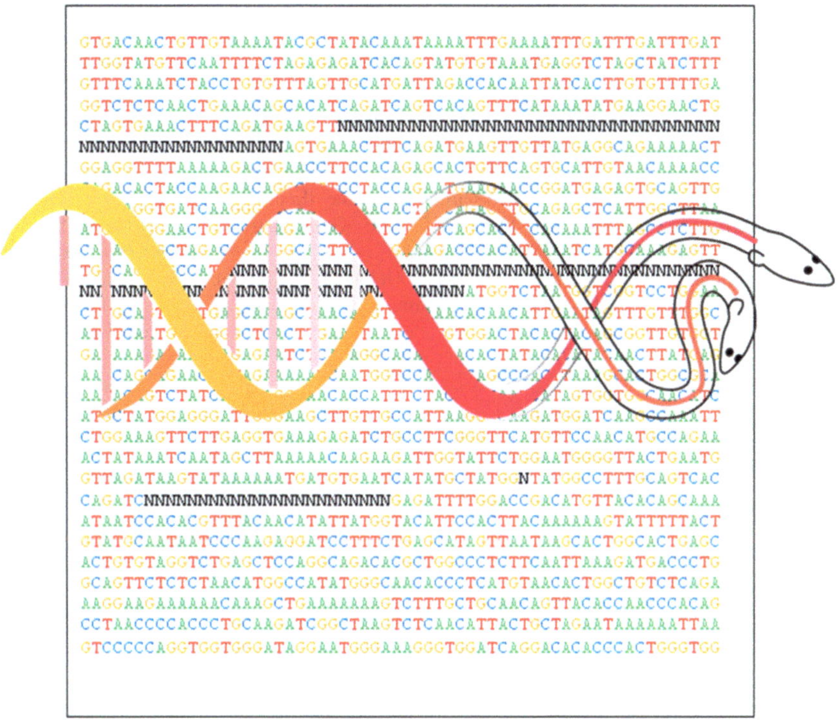

Fig. 4.20 The genomes of *Anguilla anguilla* and *A. japonica* were sequenced in January and July 2011, respectively, by ZFscreens BV, Leiden, Netherlands (http://www.zfscreens.com/home) in a consortium with K. Tsukamoto (Tokyo University), S. Dufour (MNHN, Paris), F-A. Weltzien (Oslo University), and the author

the Netherlands on eels swimming in a swimming tunnel (Burgerhout et al. 2011); the energy consumption with tags attached increased 2–3-fold. Moreover, the capacity to swim long distances decreased when tags were in place, as did peak swimming speed. Obviously, therefore, PSATs of their current size are not suitable for use in following eels right across the Atlantic.

Eels have always interested people because of the mystery surrounding them: for instance, what trigger makes them suddenly decide to migrate, exactly where do they spawn, how do the eels reach those spawning grounds, how deep can eels swim, how do eels reproduce, and how can eels be reproduced artificially. There are some new techniques that are helping to solve some of the mystery, and an important one is the sequencing of the eel genome and using that to monitor internal changes to the eel. With the genome of the European and Japanese eels now sequenced (Fig. 4.20), a major step forward has been taken, one that is crucial for the future of eel research (Henkel et al. 2012a, b).

References

Brett JR (1964) The respiratory metabolism and swimming performance of young sockeye salmon. J Fish Res Board Canada 21:1183–1226

Burgerhout E, Manabe R, Brittijn S, Aoyama J, Tsukamoto K, van den Thillart G (2011) Dramatic effect of pop-up satellite tags on eel swimming. Naturwissenschaften 98:631–634

Burgerhout E, Tunorache C, Brittijn SA, Palstra AP, Dirks RP, van den Thillart GE (2013) Schooling reduces energy consumption in swimming male European eels, *Anguilla anguilla* L. J Exp Mar Biol Ecol 448:66–71

de Graaf M, Bierman, SM (2010) De toestand van de Nederlandse aalstand en aalvisserij in 2010. IMARES Rapport C143/10: 71 pp

Henkel CV, Burgerhout E, de Wijze DL, Dirks RP, Minegishi Y, Jansen HJ, Spaink HP et al (2012a) Primitive duplicate hox clusters in the European eel's genome. PLoS One 7(2):e32231. doi:10.1371/journal.pone.0032231

Henkel CV, Dirks RP, de Wijze DL, Minegishi Y, Aoyama J, Jansen HJ, Turner B et al (2012b) First draft genome sequence of the Japanese eels, *Anguilla japonica*. Gene 511:195–201

ICES (2009) The Report of the 2008 session of the joint EIFAC/ICES working group on eels, September 2008, EIFAC occasional paper 43. ICES document CM 2009/ACOM: 15. 210 pp

Jellyman D, Tsukamoto K (2002) First use of archival transmitters to track migrating freshwater eels *Anguilla dieffenbachii* at sea. Mar Ecol Prog Ser 233:207–215

Kroes MJ, Brevé N, Vriese FT, Wanningen H, Buijse AD (2008) Nederland leeft met vismigratie. Naar een gestroomlijnde aanpak van de vismigratieproblematiek in Nederland. Rapport VA2007_33. Directoraat-Generaal Water en de Unie van Waterschappen, 93 pp http://bio.emodnet.eu/component/imis/?module=reflist&show=search@autid+64602

Minegishi Y, Aoyama J, Inoue JG, Miya M, Nishida M, Tsukamoto K (2005) Molecular phylogeny and evolution of freshwater eels genus *Anguilla* based on the whole mitochondrial genome sequences. Mol Phylogenet Evol 34:134–146

Ministry of Agriculture, Nature and Food Quality (2009) The Netherlands eel management plan. Amsterdam. 48 p http://ec.europa.eu/fisheries/marine_species/wild_species/eel/management_plans/index_en.htm

Palstra A, van den Thillart G (2010) Swimming physiology of European silver eels (*Anguilla anguilla* L.): energetic costs and effects on sexual maturation and reproduction. Fish Physiol Biochem 36:297–322

Palstra A, van Ginneken V, van den Thillart G (2008) Cost of transport and optimal swimming speeds in farmed and wild European silver eels (*Anguilla anguilla*). Comp Biochem Physiol A 151:37–44

Tesch F (2003) The eel. Blackwell, Oxford

van Dam PEJM (1997) Vissen in Veenmeren; de Sluisvisserij op Aal tussen Haarlem en Amsterdam en de Ecologische Transformatie in Rijnland, 1440–1530 (Fishing in Peat Lakes; Sluice Fishing for Eels between Haarlem and Amsterdam and the Ecological Transformation in Rijnland, 1440–1530). In: Goudriaan K, Keblusek M (eds). Verloren, Hilversum, 304 pp

van den Thillart G, van Ginneken V, Körner F, Heijmans R, van der Linden R, Gluvers A (2004) Endurance swimming of European eel. J Fish Biol 65:1–7

van den Thillart G, Palstra A, van Ginneken V (2007) Simulated migration of European silver eel; swim capacity and cost of transport. J Mar Sci Technol 15:1–16

van den Thillart G, Palstra A, van Ginneken V (2009) Energy requirements of European eel for transatlantic spawning migration. In: van den Thillart G, Dufour S, Rankin C (eds) Spawning migration of the European eel. Springer, Dordrecht, pp 179–200

van Ginneken V, van den Thillart G (2000) Eel fat stores are enough to reach the Sargasso. Nature 403:156–157

van Ginneken V, Antonissen E, Muller UK, Booms R, Eding E, Verreth J, van den Thillart G (2005) Eel migration to the Sargasso: remarkably high swimming efficiency and low energy costs. J Exp Biol 208:1329–1335

Freshwater Eels and People in France

5

Eric Feunteun and Tony Robinet

A quarter of a century ago when addressing where Japanese culture fitted in worldwide terms, Claude Lévi-Strauss pointed out that, based on the geographic dimension alone, the respective positions of France and Japan were symmetrical, because they were at either end of the Eurasian landmass. However, in terms of the cultural dimension, the two display an inverse symmetry (Lévi-Strauss 1988). The 2011 meeting in Tokyo on eels, and especially this book, gives the opportunity to a natural scientist to evaluate the comments of an anthropologist, on a subject shared by and of great interest to both French and Japanese, the eel.

To start, consider a number, in this case 10. Ten species of eel live in French freshwaters, just one in mainland France (*Anguilla anguilla*), one in French Atlantic overseas territories (*A. rostrata* in Antilles-Guyane, St Pierre et Miquelon), four in French Indian Ocean territories (Réunion–Mayotte, *A. marmorata*, *A. bicolor bicolor*, *A. mossambica*, *A. nebulosa labiata*), and five in French South Pacific territories (New Caledonia—Wallis and Futuna—French Polynesia, *A. marmorata*, *A. megastoma*, *A. obscura*, *A. australis*, *A. reinhardtii*). In this chapter the aim is to review the place of the eel in French culture, but because of the diversity of peoples and cultures in the list above and the fact that both authors are native to mainland France, focus is on the European eel there, and just a few comments are added on other eel species at the end of the chapter.

Eels in French Mainland Culture

In the late 1960s, the first author of this chapter used to vacation at his grandmother's house, which was located on the Loire estuary. Sometimes when the family gathered there, the grandmother served a dish that he, together with other children, used to hate: spaghetti with eyes! That was his first introduction to glass eels. Adults seemed to appreciate the dish at the time, but such a meal is now a thing of the past. In the 1960s, glass eels were so common in most coastal areas that they were sold at every fishmonger's stall. Generally, they were boiled and prepared as a salad along with vinegar and other condiments, or fried with garlic and parsley (Fig. 5.1). The adult eels, though, were served roasted or grilled, the flesh crushed and made into pâté, or boiled with red wine and vegetables. Anyway, glass eels in those days were considered to be a common dish, certainly not a delicacy, so common that those that could not be consumed were composted in the garden to support vegetable production. In those days too, glass eels were so abundant that they were used to make glue; they were so common that nobody really cared about them, especially about whether they would become threatened as time passed.

Nobody cared about them? Well, that is not really correct, but what is correct is that nobody in France trusted them. Eels have traditionally been considered to be snake-like sneaky creatures, so to act "like an eel" is a tag given to unreliable people who disguise themselves and shirk their responsibilities. *Il y a anguille sous roche* ("there is an eel under the rock") is another expression used to qualify unexpected situations—the river looks fine, but there are eels at the bottom to surprise you!

The origin of this aversion to eels is embedded in Latin culture. The genus and species name *Anguilla* is derived from *Anguis*, which means "snake" in Latin. *Anguis fragilis* is a lizard with no legs, known in French as *orvet* and in English as slow-worm, and in the locution *latet anguis in herba* ("a snake hides in the herbs"; Virgil). In Latin too, *anguo* means to hold tight, to strangle, to suffocate, to suffer (is there a relationship too with the English word "anger"?). In the way it was named, *Anguilla* was clearly a candidate mystical creature for ancient Latin people. This dislike was amplified centuries later, and is depicted today in front of the Notre Dame in Paris. Adam and Eve were evicted from the

E. Feunteun (✉) • T. Robinet
National Museum of Natural History, Dinard Marine Station, 38 rue du Port Blanc, 35800 Dinard, France
e-mail: eric.feunteun@mnhn.fr

K. Tsukamoto and M. Kuroki (eds.), *Eels and Humans*, Humanity and the Sea,
DOI 10.1007/978-4-431-54529-3_5, © Springer Japan 2014

Fig. 5.1 In French restaurants, glass eels are boiled and served as a salad with vinegar and other condiments, or fried with garlic and parsley. *Left* photo courtesy Jean-Francois Dareths; the other photographs are advertisements

Fig. 5.2 On the Bayeux Tapestry (1066), freshwater eels are in the river when Harold of England saved William of Normandy from drowning. © Musée de la Tapisserie de Bayeux

Garden of Eden because they trusted a snake, which suggested that Eve eat an apple from the Tree of Knowledge. That is another story, but morphological analogies with snakes render eels both mysterious and untrustworthy in Christian ideology.

When they were still abundant, everyone had a story to tell about eels. For instance, eels were reported crawling around meadows catching earthworms (which is likely a true, although rare, occurrence), someone else had an eel in their well for years, and chefs reported difficulty in slaughtering eels, which remained alive for hours even in decapitated form. These stories remind us, however, of the past abundance of the species, especially of the glass eels that formerly colonized all the estuaries of France and its Atlantic and Mediterranean coasts in southwestern Europe. Eels used to be one of the dominant species in every continental aquatic environment, from estuaries to upstream reaches of rivers. They were common in the River Couesnon when Harold of England saved William of Normandy from drowning in the eleventh century, as depicted on the Bayeux tapestry (Fig. 5.2). They were so abundant in Leman Lake (between France and Switzerland) that, in about 1277, a prelate from Lausanne cursed them. "They were so numerous that they frightened the population. People begged Guillaume de

Fig. 5.3 An ancient French tapestry first woven in the sixteenth century and reproduced in the seventeenth century showing spring activity around Pau Castle, southern France (Pays d'Orthe); a fisher pours eels from an eel pot into a fish basket. © Jean-Yves Chermeux, Musée National du Château de Pau

Champvent, bishop from 1273 to 1301, to do whatever necessary to get rid of this extraordinary mass that cruelly infested Leman Lake. Eels, despite the order, stayed. It was time to take action: the bishop summoned them to face a tribunal, but they did not present themselves, so he excommunicated them, and relegated them to a place in the lake from where they could not be so bold as to get out" (adapted from Huysecom 1999).

From ancient times until the early 1980s, the fact that understanding of their abundance, biology and ecology was shrouded in mystery made eels a popular part of human culture. In the countryside along the lower Loire, and throughout Atlantic coastal marshes, Mediterranean lagoons and the Rhône watercourse, eels were familiar to all and embedded in local culture (Fig. 5.3). Who in those days had not heard about eels moving over fields? Stories and legends are numerous, some telling of the presence of eels in some most improbable places, such as washing tanks, pools and wells.

Almost everywhere throughout its distribution, there are buildings, masterpieces, or legends that acknowledge the cultural importance and exploitation of eels. In Brittany, not far from Rennes in the town of Langon sur Vilaine, in the chapel Sainte Agathe, which is built over ancient Gallo–Roman baths, a painting dated ca. 180–200 AD shows a Venus getting out of the water surrounded by a dolphin, mounted by a Cupid and surrounded by various fish, including an eel (Fig. 5.4). In Mediterranean culture, an eel was a symbol of love and fertility, and in Morocco, the eel pool in the Merinide Necropolis of Chellah (Rabat) was supposed to bestow fertility on women who threw egg shells into it.

Fig. 5.4 *Top:* The chapel of Sainte-Agathe, Langon (Ille-et-Vilaine, Brittany) with (*bottom*) a Gallo–Roman painting on the ceiling of its cellar, representing a Venus surrounded by a dolphin and fish, among which is an eel. This painting is the only one conserved in its original place in the whole Roman western world. Photos reproduced courtesy Padrig Feunteun

Eels in Literature

Numerous classic and contemporary authors used European eels to enrich their writing. Likely the most well known is *Le Roman de Renart* (in English, "The Reynard Cycle"). This is a collection of various texts in the Roman language (ancient French), written between 1170 and 1250, showing the adventures of Reynard, a tricky and anthropogenic fox (Fig. 5.5). Reynard represents the art of speech against force and law. Those texts were among the best sellers from the twelfth to the sixteenth centuries in France. In one popular text, Reynard hides in a fisher's cart, among fish baskets, as two fishers were returning home at night. A complete translation of the ancient French text here is impossible, but in essence, seeing the two men carrying fish baskets on a cart, Reynard pretends to be dead, lying in the road, and is picked up by the men who hope to sell his fur apparel for profit. However, while lying in the cart, Reynard quietly dines on the fresh herring, then escapes with three necklaces

Fig. 5.5 A manuscript from *Le Roman de Renart*, in which a fox steals an eel from the merchant's cart. Source, Bibliothèque Nationale de France

of eels around his neck, leaving as a farewell the message: "I made with you a fair sharing: I ate your biggest herring and I take with me your best eels, but I leave the greatest in number!"

An indication of the abundance of eels in historical times is given in one of the fables of Jean de la Fontaine. *Pâté d'anguille* (eel paté) is given as an example of a delicious dish that if served too often, becomes unwelcome and less appetising. Referring then to ladies, the fable goes on to say that "even the beauty, so exquisite it is, satisfies and tires at the end. I need this bread and that bread. Diversity is my slogan" (de la Fontaine 1685).

Apart from by de la Fontaine, eels are mentioned in French literature by others. Léon Riotor wrote a book entitled "The Eel Fishermen" in 1894 (Riotor 1894), and Blaise Cendrars, a contemporary writer and traveller, said in one of his novels that he dreamt of finishing his life in the Sargasso Sea, the breeding ground of the European eel. His descendants took this statement to heart years after his death, and tried to arrange that his ashes be dispersed there. Written in Corsican in the early 1900s, Sebastianu Dalzeto's novel "Grandpa the Eel" (*Pesciu Anguilla*; Dalzeto 1930) describes the social ascendancy of a child from the misery of Bastia in 1880. The book was named by the author because of the main character's swimming ability, and also because he was able to escape from embarrassing situations. Later, Goffin (1938) wrote his "Eel Novel," published by Gallimard.

In Boris Vian's *L'écume des jours* ("The Foam of Days," Vian 1947), the translated dialogue reads:

"This eel pâté is remarkable" said Chick. "Who gave you the idea to make it?"

"Nicolas had the idea" said Colin. "There is an eel – there was – that came into his washbasin daily through its cold water pipe."

"That is curious!" said Chick. "Why did it do this?"

"It stuck its head through the tap and emptied the toothpaste tube by biting a hole in it. Nicolas only uses American paste with pineapple flavour, and the eel obviously wanted it."

"How did he catch the eel then?" asked Chick.

"He put an entire pineapple on the basin instead of the tube. When the eel used to eat the paste, it could swallow it and then turn its head back, but with the pineapple, it could not do that, the more it pulled, the more its teeth were embedded in the pineapple. Nicolas …" Colin stopped.

"Nicolas what?" said Chick.

"I hesitate to tell you, it may deter your appetite."

"Come on" said Chick, "I don't have that much appetite left!"

"Nicolas came into the room at this time and cut off the head of the eel with a razor blade. Then he turned on the tap and all the rest of the body came out of the tap."

"That's all?" asked Chick. "Give me some more pâté. I hope that the eel has a large family in the pipe."

"Nicolas put some raspberry paste in the tap to see…" said Colin.

Many popular French expressions and proverbs are based on eels, and many are defamatory about the fish. *Être comme une anguille* is "to slip out like an eel," describing someone who is difficult to pin down when needed. *Il y a anguille sous roche* directly translates as "there is an eel under the rock" (as mentioned earlier), i.e. something is about to happen but is hidden; something fishy is going on. *Filer comme une anguille* means to escape and disappear in order to avoid facing an embarrassing situation. These proverbs are popular because eels used to be abundant and are a part of French collective memory.

Six Centuries of Eel Science in France

There is an extensive technical and scientific literature on eels in France, including articles in fisheries or other scientific journals, monographs and scientific essays. Among them, the monograph on eels written by Léon Bertin of the French National Museum of Natural History in 1942, and the more recent book by Yves-Alain Fontaine (*Les anguilles et les hommes*, "Eels and Humans"; Fontaine 2001) are well known in the French eel community. More recently, *Le rêve*

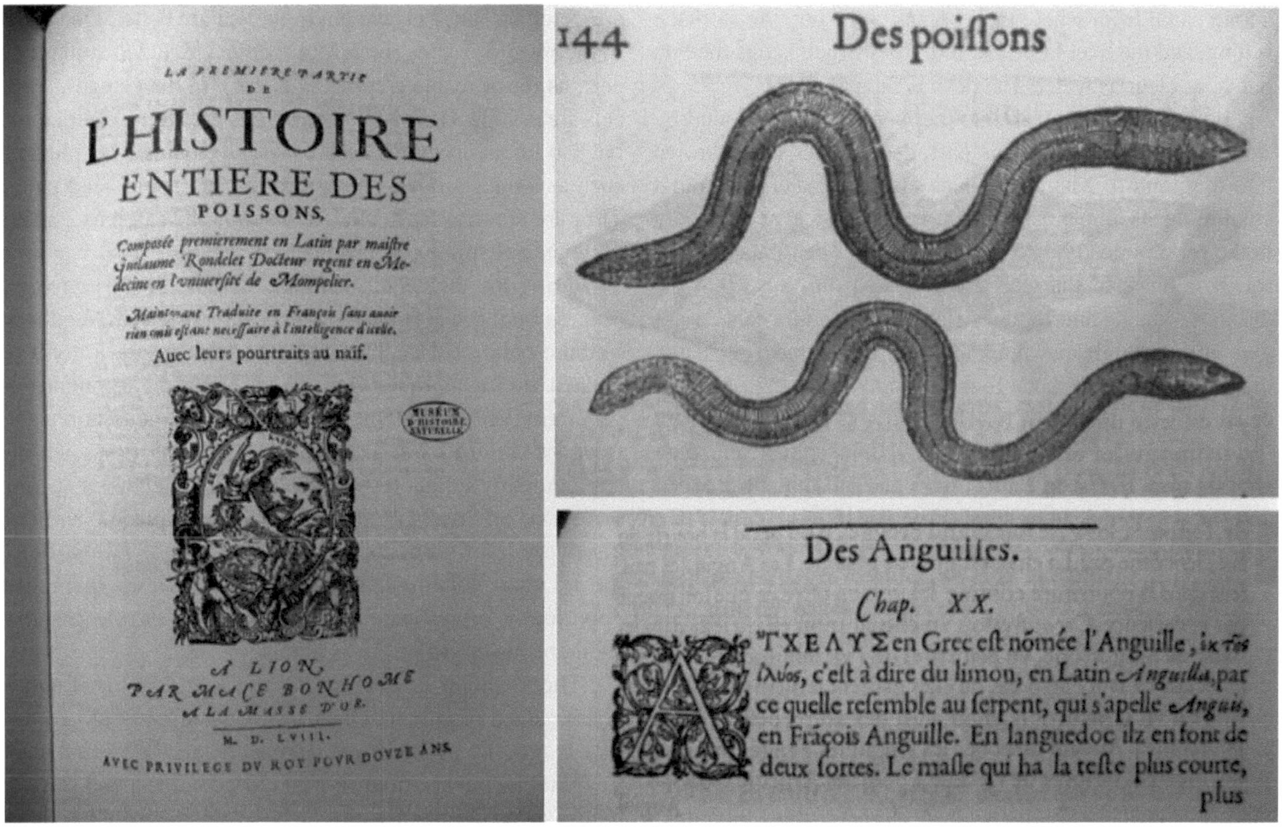

Fig. 5.6 The cover (*left*) and some pages from *L'Histoire Entiere des Poissons* ("The Full History of Fish") by Guillaume Rondelet (1558), translated from Latin to French by Laurent Joubert and recently re-edited by François Meunier

de l'anguille ("The Eel's Dream"; Feunteun 2012) provides an update on eel biology. These books, as do the many other publications on eels not mentioned here, show how interested European scientists, and French scientists in particular, are in the species. Below, therefore, an attempt is made to provide a non-exhaustive overview of some of the works of perhaps the best known French scientists who have worked on eels.

The earliest record of French scientific work on the biology of eels is that of Guillaume Rondelet (*Gulielmi Rondeletii*; Rondeletti, 1554), a Doctor regent of the University of Montpellier, who wrote in Latin his *Libri de piscibus marinis in quibus verae piscium effigies expressae sunt*. It was translated into sixteenth century French by Laurent Joubert, as *L'histoire entière des poissons* ("The Entire Story of Fish"; Rondelet, 1558; Fig. 5.6). In the second part of that treatise, a large part (Chapter 20 of the Book on River Fishes) is devoted to considerations on the biology of the freshwater eel. Rondelet recalls, *inter alia*, Aristotle's statement about the mysteries of eel reproduction, because no egg had ever been found, and reminds the reader of the suggestions of Pline and other philosophers. He also gives his views on eel reproduction, perhaps the first scientist to do so. In spite of the adult eels in freshwater not having visible gonads, he suggests that they are hidden by fat. In the extract below, translated into English by the authors for the purpose of this chapter, from the original text in ancient French of Joubert, re-edited and commented upon by François Meunier in 2002 (Rondelet 2002), he cites ancient proverbs on eels, and outlines his personal preferences regarding eel cuisine.

"*ΑΓΧΕΛΥΣ* in Greek is named the eel, *εχτηςιλυος*, which means "from silt," in Latin *Anguilla*, because it looks like a snake, which is named *Anguis*, in French *Anguille*. In Languedoc, southern France, there are two types: the male, which has a shorter but larger head, is referred to as *margaignon*; the female, which has a smaller, more-pointed head, is referred to as *anguille fine* ("thin eel"). [...] Each eel gives birth in freshwater, and it is the only one of all similar fish that enters marine lakes, including the sea, but otherwise lives in rivers, lakes and ponds. It is a long, slippery and scaleless fish, covered by leather that is easy to pull off. It has a large mouth, with tiny teeth and tiny ears, the latter covered by skin but with a small opening that causes them to suffer asphyxia in cloudy waters, which is why they can live quite a long time out of water. Close to the ears are two very small wings. By bowing the body they move in water. Instead of wings, from the middle of the back, and from the anus, they have an edge, more skin than wing, that surrounds the rest of the body. They have a long canal from where their organs come out, the stomach is long, and the liver large and red, from which hangs the bile purse, as clear as water. The flesh is fat and slimy.

"Eels give birth in rot, like the worms in earth found by experience. Because in days past, dead horses were thrown into the pond of Maguelone and sometime later one found countless eels, one must not believe that they give birth only in the rot of a dead horse, but also in other beasts, and other rot. It is not said that eels are created from those that die through old age, and rotten. Aristotle writes that eels do not give birth through spawning, that they do not have eggs, and that none have ever been found with eggs or seed, nor any canal for seed or for matrix. Therefore this type of fish, with blood, gives birth without eggs and without spawn, and what we do know is that in each silted pond, all the silt dredged and thrown on the shore gives rise again to more eels if rain falls, whereas during dry periods they cannot give birth, because they require the rain to feed them. In eels, therefore, there is no difference between males and females. For this reason, the differences mentioned previously concerning their heads will make a species difference, not a sex difference. Pliny had another opinion on eel regeneration. Eels, he says, rub against rocks and the scum so scraped off comes to life; there is no means of their generation. Athenaeus wrote that eels spawn by fighting with each other, from which fight comes the scum, or slimy humour, from which when fallen in silt, young eels are born. Oppian writes that he had seen eels embracing each other and spawning, and thinks that all parts needed for their generation are not lacking, because below the belly females do have a canal for the matrix, males a canal for seed. However, these parts seem to be covered with fat, as eggs are for the same reason. Therefore, there are eels that give birth by spawning between males and females, and others that give birth in rot.

"Eels live in fresh and clear water, and for this reason, those who construct eel ponds, as Aristotle wrote, should be careful that in places where eels are contained, only clear water runs, because if the water is cloudy, they suffocate. That is why in rivers of France during winter with during extensive rain or melting snow, water levels increase and become cloudy, so in trapnets and other nets able to catch fish, one catches large quantities of eels, which are salted for the next Lent. For this reason Aristophanes compares those who make profit from wars and perturbations of republics with eel fishers, who catch little if nothing clouds the water. From this, according to Aristophanes, comes the common proverb "to fish in cloudy waters," i.e. to catch an eel one must make the water muddy, which is expounded by those who make profit from wars, mutinies, quarrels, and trials.

"Eels do not float on water to be seen, and rarely come to the surface like other fish, because they have a small belly and therefore a little air inside can sustain them. One should salt eels to make them taste better, because their viscosity is corrected by salt. Eels caught in the sea are better. One catch in the pond of Lattes (near Montpellier) included some long eels, as much as three cubits (135 cm). In the Ganges, some eels are as long as 30 ft (978 cm). Eels are slimy, nourishing, but bad for any person subject to disease arising from phlegm, such as kidney stones, gouty arthritis and some head pains. They are not healthy in pâté, and should be eaten as starters at table. Spit-roast eels taste better."

During the eighteenth century, Voltaire (1785) gave his opinion on the secret birth of eels, as reported by Needham. Voltaire believed Needham to be an Irish Jesuit, a myth he created during a long controversy regarding spontaneous generation of eels in flour, referred to as "the generation controversy." In 1743, Needham had published his first scientific paper, mainly on geology. However, he appended a section on "some microscopic discoveries I lately made," and one discovery led to him forever being known as *l'anguillard*, the eel man. Voltaire attacked Needham's views, taking the opportunity to castigate the latter's views on the possible spontaneous origin of life as atheistic:

"There was in France around the year 1750 an English Jesuit named Needham. Disguised in secular [monk] … this man was doing physical and mostly chemical experiments.

"After he put rye flour in well closed bottles, and juice of boiled sheep in other bottles, he believed that from his sheep juice and his rye were born eels, these latter producing themselves others; and that so a variety of eels was formed indifferently in a meat juice, or a rye grain.

"A physicist with reputation did not doubt that this Needham was not a deep atheist. He concluded that because one makes eels with rye flour, one could make men with wheat flour, that nature and chemistry produce everything, and that it was demonstrated that a God who creates everything is useless.

"… it is so strange that men, to deny a creator, have attributed to themselves the power of creating eels.

"… Not a long time ago one assumed that in Brussels a rabbit made half a dozen of young rabbits to a hen. This transmutation of flour and sheep juice in eels was evidenced to be as false and ridiculous as it was in fact, by M. Spalanzani, a bit better observer than Needham. His observations were not even necessary to demonstrate the extravagance of such a palpable illusion. Soon the eels of Needham were going to meet the hen of Brussels."

From the time of Voltaire to the twentieth century, numerous respected scientists gave opinions on the mystery of the eel life cycle. Le Comte de Buffon in his *Histoire Naturelle* (Comte de Buffon 1828) stated "Eels that form in the glue made with flour do not have any other origin than the meeting of organic molecules of the most substantial part of the grain: the first eels that come out are certainly not produced by other eels; despite the fact that they have not been spawned, they can spawn themselves some other living eels.

One can by cutting them with a sharp lancet see the small eels coming out of their body, and even in large numbers, it seems that the body of this animal is only a scabbard or a bag that contains a large number of other small animals, that are maybe themselves only the scabbard of the same species, in which, by growing, the organic matter is assimilated and takes the same shape as eels." Unfortunately, there is insufficient space here to reproduce an exhaustive overview of the abundant literature on eels produced in France during the nineteenth century.

Much later, in the 1940s, some years after the Italian Grassi had shown that *Leptocephalus brevirostris* was actually the larva of an anguillid eel (Grassi and Calandruccio 1897a, b), and nearly a quarter of a century after the Dane Johannes Schmidt proposed his Sargasso Sea hypothesis of spawning, Léon Bertin worked on the biology of eels at the French Museum (Bertin 1942). One of his students, Maurice Fontaine, was among the first to understand that all eels living in inland habitats were juveniles, or adolescents, and later offered some memorable insights into fish physiology, showing that release of the pituitary growth hormone was triggered by environmental parameters such as pressure and temperature. This discovery started a long tradition of research in France into the physiology of eel reproduction, now led by Sylvie Dufour at the French Museum of Natural History in Paris, who recently coordinated, along with Dutch counterpart Guido van den Thillart, a notable European project on eel reproduction, called EELREP.

In the 1970s, Pierre Elie of CEMAGREF (the French Research Centre on Agriculture and the Aquatic Environment) also initiated a long research programme investigating the close relationships between humankind and eels. He was among the first to study the estuarine behaviour of glass eels, and proposed a key for determining the age of glass eels based on colour patterns on the head, tail and body. He was also one of the first to raise concern in the 1980s when the decline in the numbers of arrivals of glass eels became so obvious. His work triggered more effort in implementing management plans in France and other countries around the Mediterranean Sea.

In the 1980s, Gérard Marquet, an independent researcher who successively held teaching posts at high schools in Tahiti, New Caledonia, Mayotte and Réunion, was the first French national to work on the ecology of tropical eels in French overseas territories. He produced a thesis on the community and population structure and levels of exploitation of eels in French Polynesia in 1987. It was he, therefore, who was responsible for initiating the research effort on tropical eels that French scientists continue until today.

In the 1990s, Raymonde Lecomte-Finiger from the CNRS, the French National Research Centre in Perpignan, and Yves Desaunay from Ifremer, l'Institut français pour l'Exploitation de la Mer, in Nantes, examined the otoliths of glass eels captured around the coasts of Europe. They inferred from the number of daily rings they counted in >1,000 glass eel otoliths that the transoceanic migration of glass eels lasted no more than a year, in direct contrast to the then more than 80 years of understanding based on the work of Johannes Schmidt (see the Danish chapter elsewhere in this book), who stated in 1922 that the transoceanic larval migration of the European eel took 3 years. That French finding is still the subject of controversy and heated debate 20 years on, recent papers based on ocean current transport mechanisms still supporting the 3-year transoceanic movement of leptocephalus larvae to Europe. Perhaps the truth is somewhere in between?

The French Museum of Natural History now concentrates its efforts on the ecology of tropical eels in the Indian Ocean, and on the marine stages of European eels in the Atlantic, supported by long-term surveys of recruitment into small (Frémur and Oir) and large rivers (Loire). With this work, the Museum has contributed to the so-called EELIAD project (see the UK chapter) of the EU that aimed to enhance understanding of the mystery of the marine migration of silver eels. Given their knowledge of the physiology and ecology of the European eel, French scientists were involved in the project in evaluating the relationship between the quality of eels and breeding success.

In the 1980s, at most ten researchers and students focused on eel science in France, but today that number has risen to about 30.

Eel Fishing in France

In France, at least until the downturn and the initiation of recovery plans (management plans), eels were fished at all life stages. When they enter western French estuaries, the glass eels that have metamorphosed from the leptocephalus larvae are harvested by dipnets (Fig. 5.7); depending on site, they are caught from the riverbank or from a small boat. Professional fishers also use pushnets operated at 2 or 3 knots in the lower reaches of estuaries. In northern France, the pushnets are 1.2 m in diameter and fished 2–3 m deep, whereas in the Gironde estuary, and more generally in southwestern France, fishers use two rectangular dipnets with a total surface area of 14 m^2.

In the 1980s, more than 3,000 professional and many thousands more amateur fishers fished for eels. At that time, the market value of glass eels used to rank as first or second of any species taken by the fisheries in the Bay of Biscay and its local estuaries. In the Loire and Vilaine rivers, respectively, 770 and 106 t of glass eels were caught in 1977, but today there are just 661 fishers still taking glass eels in France, Loire catches had dropped to 21 t by 2007 and Vilaine catches had declined to 3 t by 2010. Moreover, the export ban on glass

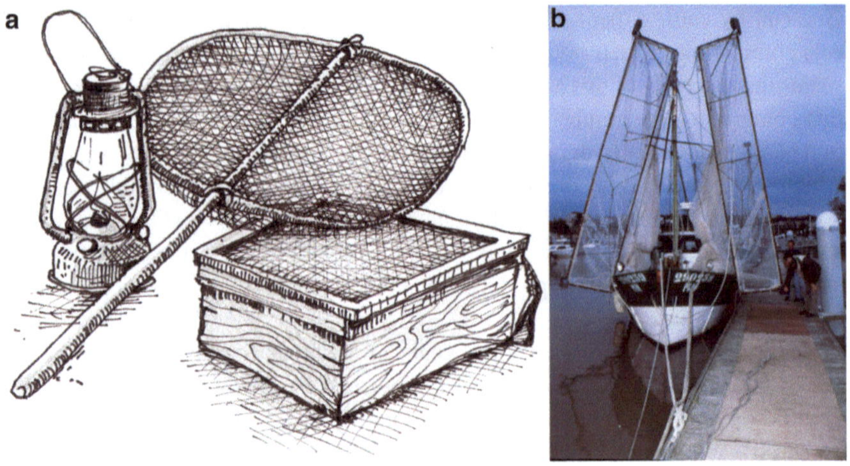

Fig. 5.7 (**a**) Traditional fishing gear used by single fishers in France to catch glass eels. (**b**) There were as many as 3,000 professional fishers in the 1980s, but fewer than 600 operate now. Source, Jean Louis Darrière and EF, reproduced with permission

eels recently imposed by CITES (the Convention on International Trade in Endangered Species of Wild Fauna and Flora) collapsed the market price for glass eels and the number of fishers will therefore continue to decline. On the other hand, of five million amateur fishers in France, half catch an average of five eels per year, a total of ten million eels!

The ways to catch yellow and silver eels in France are more diverse and tend to be specific to each region. One of the most common means, however, was *la pêche à la vermée*, i.e. to take a thread of wool, pull it through a number of earthworms to create a "worm-thread" 1–2 m long, bundle this into a knot and tie it to a rod, then dip this into the river (known in some countries as bobbing). Eels were attracted to this bait and bit it, getting their teeth stuck, their struggle to escape was felt on the rod, and they could be pulled out of the water. In France it used to be common to hold an umbrella upside down (Fig. 5.8) under the bait so that the eel dropped off the bait to be captured. This technique was reported to be very efficient, allowing dozens of eels to be captured within a few hours, so was banned in the late 1980s as eel numbers dropped.

Also, as elsewhere in the world, wicker pots were used for centuries in France to catch eels, although from the 1960s they were largely replaced everywhere by more-efficient fykenets. Nets were also commonly set at the outlets of watermills, mainly to catch silver eels (Fig. 5.9), and *tendins* were used in reclaimed marshes where the right to use nets was decided by public auction. Similar traps were common throughout France (Fig. 5.10), and tradition has it that many millers used to use the proceeds of their eel catches to support maintenance and replacement work in the mills.

In September 1988 in the banked marshes of Bourgneuf-en-Retz in western France, we helped local farmers clear the mud from an ancient saltpan that had been converted to support fish farming. It was hard work because it was being done in the traditional way, using small shovels to remove the mud onto the surrounding meadows (Figs. 5.11 and 5.12). Eels were soon encountered though, seized with an "eel scissors" and dropped into a basket, then subsequently relocated to a holding cage in a neighbouring pond. Although the work was hard, it was rewarding because, after just 2–3 days of work, a layer of silt 20 cm thick had been removed from the 2,000 m^2 saltpan, and some 200 kg of eels had been collected. Although the catch was impressive then, it was reported that three times as many eels were taken in the 1970s from the same waterbody! Every 3–5 years, such pans, filled with brackish water from the sea through a series of sluice gates (preventing seawater intrusion inland) had to be cleared of the silt and mud deposited by the sea and occasionally by collapsing banks. Up to the 1980s, the catches of eels associated with such clearance activities were well worth the effort and justified the maintenance of the ditch network throughout western France: >40,000 km of ditches covering >200,000 ha of reclaimed saltmarsh along the French coast of the Bay of Biscay alone. Such habitats were considered capable of supporting as much as 2,000 t of eels, including 400 t of silver eels, and fishers were thought to be taking half of this production annually, allowing some 200 t of silver eels to escape to the sea annually.

The maintenance of these ditches and saltpans, which originated in the eleventh century from the activities of monks producing salt, was for centuries carried out using shovels, a heritage across centuries that formed a sort of symbiosis between man, eels and water. The subtle balance created a unique landscape that sheltered a rich diversity of fish, birds and plants that has since been recognized as a rich heritage of high conservation value. However, the decline of the eel and the pressure to gain land suitable for agriculture eventually led to the areas being converted into arable land, destroying productive eel habitats, and demonstrating clearly how a human society with an economy based partly on an eel fishery can create, maintain and ultimately destroy a rich and diverse aquatic landscape. It took <1,000 years for the rich aquatic landscape to be created and destroyed, and now there

Fig. 5.8 (**a–d**) Montage of a well-known means of catching eels, *la pêche à la vermée* (see text for description). Source, Outback Images (www.outbackimages.fr), reproduced with permission

are virtually no eels, no fishers, a much poorer bird fauna (with no herons) and a concomitant loss of biodiversity.

Silver eels can be caught with practically any of the gear described above, but only (in France) between November and January. They start migrating back to the Sargasso Sea during dark, stormy and wet nights, often a strong river swell heralding the start of the migration. In Mediterranean lagoons, fishers move their pots or fykenets (*capetchades*) to the outflows in an attempt to intercept the silver eels at the start of their migration (Fig. 5.13), and in rivers along the Atlantic and Channel coasts of France, nets used to be set on the main watercourse to catch the eels during winter floods (Fig. 5.14). France used to have many fisheries for silver eels, but many were closed not only because of the decline in the number of eels, but also because of contamination of waterbodies by PCBs (polychlorinated biphenyls). The last fishery remaining is in the downstream reaches of the Loire, and that has been in existence for centuries. Originally in the Loire fishery, at least 300 fishers operated pots at the end of a willow fence 100 m long that channelled silver eels towards the pots. One of the difficulties originally associated with that fishery was that the fence was easily shifted by the strong flood currents, so in the 1950s, new gear was imported from the Netherlands, an anchored stownet, which is a net 50 m long with an opening of 50 m^2 at the correct depth for catching eels, along with a winch (Fig. 5.15). In all, 14 fisheries for silver eels existed in France up to 2000, but now there are just eight, likely because retiring fishers were not being replaced by younger fishers, most of whom considered the future of the fishery to be uncertain.

Fig. 5.9 An eel weir in a western France marshland. Photograph courtesy IRSTEA Bordeaux, reproduced with permission

During the second half of the twentieth century, the total annual Loire fishery catch of silver eels amounted to 50 t of large female silver eels, each weighing an average of 0.8–1 kg, but occasionally up to 4 kg. More recently and associated with the declining effort, the annual French catch dropped to <25 t, but the catch rate seems to be remaining fairly stable year on year. Between 2001 and 2008, a mark–recapture experiment revealed that the stownet fishery of the Loire alone caught ~15 % of the eel run annually, and an estimated 300,000–350,000 silver eels (300 t) escaped from the Loire to the sea annually (Adam et al. 2008). The exemplary collaborative work involving fishers and scientists that resulted in those values underpinned the first reliable estimate of the production of silver eels by a large European river system, allowing the fishers in effect to become partners in the management of eels in France.

These descriptions are just a few examples of the diversity of fishing techniques, gears and fishers exploiting eels in

Fig. 5.10 Eel traps known as *Braies* on the River Vendée, western France. Photographs courtesy Antoine Legault, Fish-Pass, reproduced with permission

Fig. 5.11 Clearing a saltpan converted to fish production in the reclaimed marsh of Bourgneuf en Retz (western France), using shovels. *Top*: A breach is opened with a shovel to empty the saltpan, and a wooden water board planted in the mud in front of an eel basket, to move the remaining water. *Bottom*: Once free of water, the mud is cleared from the saltpan with shovels, and the eels caught with an eel scissors and placed in the eel basket. Photographs courtesy IRSTEA Bordeaux, reproduced with permission

Fig. 5.12 Traditional tools used in the reclaimed saltmarshes of western France: a water shovel to push the stagnant water out of the saltpans; a mud shovel to clear the saltpans; an eel scissors, with dulled blades, used to catch the eels and place them in an eel basket; a wooden holding box floated in a pond until the eels are sold or consumed; an eel pot with a wooden frame, which is placed in gliders of a fishing weir. Photograph courtesy IRSTEA Bordeaux, reproduced with permission

France, in rivers, lagoons and other aquatic ecosystems. As they operate, fishers observe and report on the waterbody and its status, then have to wage active campaigns to preserve the ecosystems on which the eels, and hence their livelihoods, depend. The role of the fisher in maintaining stocks and diversity is obviously crucial; whenever a waterbody use project is being considered, it is not only the fish but also the fisher that is concerned. Succinctly, if there are no fish, there will be no fishers.

Chronicle of a Decline

The decline of the European eel is described and explained in many ways in this book, and in the published literature generally. There is no doubt that the species is in serious decline and has been on a downward trend since at least the 1980s, and almost certainly before that. Catches of glass eels are as little as 1 % of historically recorded catches, and natural recruitment has been so depressed relative to historical levels that Baltic Sea states initiated restocking efforts as long ago as the 1960s, Germany even earlier. Yellow and silver eel stocks are seemingly in a slightly better state, however, given that catches in Europe are currently at some 30 % of historical levels.

The likely causes of the decline have been documented scientifically and in the popular press for decades, including by a French Working Group on Eels in 1984 and subsequently by successive ICES Working Groups, and are unchanged and agreed by the international community. In reality, though, even if the likely causes are known, the decline is not that well understood, scientists generally agreeing that numerous factors are acting in synergy. Global climate change, for instance, may well have caused changes in ocean trophic webs, reducing the productivity of oceans, and suggestions have been made by a few that the Gulf Stream is being diverted north in concert with the warming. As a consequence, the leptocephalus larvae that originate in the Sargasso Sea may be growing

Fig. 5.13 An eel fykenet (known locally as a *capetchade*) used in the coastal lagoons of southern France. Photographs courtesy F. Charrier, Fish-Pass, reproduced with permission

more slowly, having to migrate for longer periods of time, and hence be subjected to increased mortality.

The same global climate change may also be responsible for the extensive modification in some inland habitats. The land areas covered by large rivers such as the Rhône, the Gironde, the Loire and the Seine, have been reduced to as little as 10 % of their original state by dredging, embankment and reclamation. Considerable aquatic habitat area has also been converted to agriculture or urbanized. During the late nineteenth century too, many thousands of dams were built on rivers and there was no law forcing developers to mitigate their effect on fish movement, including eel migration upstream and downstream. As a consequence, many rivers have no eels in their upstream reaches.

Eels grow throughout their life until their return migration to the Sargasso Sea, when they stop feeding, storing fat to support their single spawning in the Sargasso Sea. Many contaminants are stored in fat and are not released until spawning, and contamination by metals modifies fish behaviour by reducing orientation ability or impeding reproduction. Many persistent organic pollutants (POPs) trigger early silvering and related migration by eels, before the fat stores are sufficient to fuel the transoceanic migration. Other pollutants are released during vitellogenesis and almost certainly reduce the survival likelihood of eggs and larvae.

Direct mortality through turbines may kill many or even all silver eels during their downstream run. For example, eels in the Rhône River departing from Lyon, 250 km upstream

Fig. 5.14 *Anguillères* (eel gears) used in the Bay of Somme, northern France. Photographs courtesy Antoine Legault, Fish-Pass, reproduced with permission

Fig. 5.15 *Top:* A boat carrying a stownet on the River Loire and (*bottom*) a drawing of how the stownet is anchored. Photograph EF, drawing courtesy Conseil Supérieur de La Pêche, reproduced with permission

of the delta, have to pass through 11 hydroelectric works on their way to the sea. If each turbine accounts for 10 % mortality of the eels passing, then the probability of survival into the Mediterranean Sea is close to 30 %. The situation is the same in many large rivers, but also in the many small rivers that are suitable habitat for eels. Mortality of eels also eventuates from abstraction of freshwater for agriculture, industrial use and cooling of hydrothermal plants.

Fisheries were for many years considered the major cause of the decline in the population of the European eel, but nowadays it is generally admitted that even if the eel fishery accelerated the decline of eels, blame for the collapse cannot be laid solely at the door of fishers.

A Management Plan for Eels in France

It is difficult to create and implement an efficient management plan in a country where belief has always been that the eel resource is for some mysterious reason inexhaustible. Up to 1984, the year the French Working Group on Eels alerted the Ministry of Environment that glass eel catches were declining rapidly, the eel was still considered to be a pest in France. Glass eels were fished with absolutely no restriction, and French law actively promoted their destruction when eels were encountered in rivers also containing salmonids (salmon and trout). Despite the warning of the decline being strongly worded and supported by ICES, it was not until 1998 that ICES formally recorded that "the [eel] is outside safe biological limits" and that it was necessary to "reduce fishing mortality as much as possible". A year later, the same ICES Working Group modified its text by proposing "reducing all anthropogenic mortality causes" to a level as close to zero as possible. This was the first implicit recognition that multiple factors had led to the collapse of the population. Six years later, in 2005, the ICES working group stated that "it is of [great urgency] to implement a management plan aiming to enable the escapement of 40 % of silver eels that would be produced by rivers in pristine conditions". Then, in 2007, the European eel was listed in Annex 2 of CITES, and the same year, the European Commission formally decided that each Member State should implement a national management plan for the European eel.

In 2008, France implemented its eel management plan, based on four main actions: a reduction in fishing mortality (30 % reduction by 2012), a reduction in turbine mortality (no threshold mentioned), an improvement in fish passage through barriers, and a restocking plan. However, the management plan does not aim directly to restore water quality or surface area because these objectives are considered covered by the EU's Water Framework Directive.

After the CITES Conference of Parties voted to list the European eel in its Appendix II in 2007, the scientific review group in Brussels decided in 2008 to ban trade in glass eels outside the European Union progressively (2008–2010). That decision caused the market price of glass eels to collapse and summarily halted the French fishery for that form of eel, rendering the national restocking project unfeasible in 2011. The management plan had aimed to restock 4.4 t of glass eels annually over the western coasts of France, but because of the low market price, many fisheries for glass eels simply stopped operating in 2011 until the government offered to provide the funding necessary to support them into April. By April, though, the glass eel run was over in most rivers, and catches were poor, so ultimately <800 kg of glass eels were restocked, just 1.7 % of national catches rather than the 30 % announced initially. In 2012, however, the restocking plan worked reasonably well and the objective to restock 4 t was achieved.

Tropical Eels

Away from the French mainland, in Polynesia, in Mayotte, in New Caledonia and in Saint-Pierre et Miquelon (North America), mythology and legend places the eel at the heart of human culture. In Madagascar and the Comoros Islands, depending on ethnic group and even on village in some instances, eels are either taboo (forbidden) or much appreciated. It is forbidden to fish in some villages in Madagascar unless a spell is cast by applying a special ritual. For instance, glass eels along the Pangalanes Canal on the east coast of Madagascar are taboo, but they do have a certain economic attraction, a consequence of speculation on tropical eel fisheries as a result of the collapse of the market for European eels. The result has been that male fishers are still not permitted to catch eels, but female fishers are allowed to catch eels there only at night and if dressed in their birthday suit!

In Tahiti, there is the story of Princess Hina, the daughter of the Sun and the Moon, who was so beautiful that her body, especially her hair, shone with light rays. She was promised to King Vaihiria, an enormous eel, but Hina escaped and found protection with an old fisher in the lagoon, the great God Maui. However, King Vaihiria soon found her, because at night he could see her hair shining through the walls of any house. He captured her, but Maui found him and cut off his head, which rolled to the feet of Hina. The head then whispered to her "one day will come when everyone who hates me will kiss me on the lips, and you will be the first of them". Maui wrapped the head in banana leaves and gave it to Hina, saying to her "you must bury the head in your village, but do not place it on the ground before you have arrived at the village or the curse could become reality." On the way, the heat was so great that Hina had to put the bundle on the floor in order to drink from a clear river. Suddenly the ground opened and swallowed the head, and a strange plant sprang up and began to grow. It looked like an eel and its head rose to the sun: it was the first coconut palm. Later there was a terrible drought, water sources ran dry, all trees except the coconut palm died, and Hina, thirsty, put her lips to the holes of a nut that looked like two eyes and a mouth, and drank the milk inside. She then heard a voice coming from nowhere saying "you will be the first of them."

Polynesian people tell us with a smile that once one has heard this story, one cannot look at a coconut in the same way (a coconut, viewed from above with its three dark spots, looks like an eel head with two eyes and a mouth). In any case, the comparison with the coconut palm clearly demonstrates the importance of eels in human culture, and even as a good food dish, for Polynesians.

Conclusion

For centuries, eels have been an important fish and part of French culture, beliefs and expression. They are symbols of evil, depict the mystery of life's origin and its complex relationship to the water cycle, and are symbols of abundance and fertility. As highly migratory species, eels also intuitively represent freedom to move in an open water system from the sources of rivers to the deep ocean. This freedom is being strongly limited by anthropogenic activity that modifies and degrades the land–river–ocean continuum so essential to the biological cycle of eels. Hopefully, though, the eel will not become a mere symbol of how humankind should live in harmony with nature. Indeed, whereas unprecedented effort has been made in attempting to restore eel populations worldwide, they keep declining sharply, perhaps exemplifying man's inability to manipulate nature when faced with many conflicting objectives and needs.

The mystery of the biology of the eel still remains a challenge for European society to address, and the aim of protecting and saving the species ultimately depends on whether future research can solve the manifold unknowns, in particular the marine phase of life and to control or stimulate reproduction. It is also crucial that the cultural value of the species be communicated and appreciated; we must not allow the large-scale extinction that befell dinosaurs to happen to eels. Their ongoing survival for future generations of humankind to know them in live form is surely important given that they have already survived massive environmental upheavals over millennia, including glaciations, global warming and current shifts.

References

Adam G, Feunteun E, Prouzet P, Rigaud C (2008) Guide Méthodologique pour la Définition d'Indicateurs d'Abondance, de Colonisation, d'Exploitation de l'Anguille Européenne (*Anguilla anguilla*) et de la Qualité de ses Habitats. Editions QUAE, Paris, 393 pp

Bertin L (1942) Les Anguilles. Bibliothèque scientifique, Paris
Comte de Buffon (Leclerc, G-L.) (1828) Histoire Naturelle. 12. Histoire des Animaux. Chapter 9. Variété dans la Génération des Animaux. p 364
Dalzeto S (1930) Pesciu Anguilla. Notre Maquis, Paris
De La Fontaine J (1685) Pâté d'Anguille. Contes et Nouvelles en Vers. Henry Desbordes, Amsterdam
Feunteun E (2012) Le Rêve de l'Anguille. Un Sentinelle en Danger. Petite Encyclopédie sur un Poisson extraordinaire. Buchet Chastel, Paris
Fontaine Y-A (2001) Les Anguilles et les Hommes. Odile Jacob, Paris
Goffin R (1938) Le Roman des Anguilles. Gallimard, Paris
Grassi GB, Calandruccio S (1897a) Riproduzione e metamorfosi delle anguille. Pesca Acquacultura 7:193–202
Grassi GB, Calandruccio S (1897b) Riproduzione e metamorfosi delle anguille. Pesca Acquacultura 8:225–233
Huysecom R (1999) Mille ans de Pêche au Léman: des Hommes, un Lac, un Métier. Editions La Salévienne, Paris, 158 pp
Lévi-Strauss C (1988) Place de la culture japonaise dans le monde. Presentation made on 9 March 1988 in Kyoto, during the opening ceremony of the International Research Centre on Japanese Studies (Nichibunken). Article first published in the Journal *Chuô kôron, Tôkyô* in May 1988 (in Japanese)
Riotor L (1894) Le Pêcheur d'Anguilles, l'Éternelle histoire, Poèmes Légendaires. De la Plume, Paris
Rondelet G (2002) L'Histoire Entière des Poissons, Foreword by François Meunier and Jean-Loup d'Hondt. CTHS, Paris [facsimile of original work held in Lyon, by Macé Bonhomme, 1568]
Vian B (1947) L'Écume des Jours. Gallimard, Paris
Voltaire (1785) Histoire des Anguilles sur Lesquelles est Basé le Système. Oeuvres Complètes, vol 19. Garnier and Louis Molland, Paris

6 Eels and the Japanese: An Inseparable, Long-Standing Relationship

Mari Kuroki, Martien J.P. van Oijen, and Katsumi Tsukamoto

The Japanese nation has a long history of using freshwater eels as food, and these days up to ~100,000 t of eels per year are consumed there, about 70 % of the world's eel consumption. In this context, there is no doubt that Japanese have a closer relationship with eels than any other nation in the world, and therefore need to take a heavy responsibility for their conservation.

The well-known eel dish *kabayaki* (filleted eels dipped in a sweet soy-based sauce then broiled on a grill) was invented in the middle of the Edo era (1603–1868), more than 250 years ago, following an increase in soy sauce production around Edo (Tokyo). Increases in eel consumption in those days yielded a variety of objects related to eel food culture, such as eel dishes, cooking knives and tableware. The food culture of eels expanded and deepened into idioms and folklore, arts and crafts revolving around eels, and further, to beliefs and spirituality related to eels. To know the cultural aspects of eels is to understand the relationship between eels and humans. This in turn may elicit an awareness of conserving eels, the populations of which are now in drastic decline worldwide (Kuroki and Tsukamoto 2012).

In this chapter, we briefly overview the historical progress of the science of eels in Japan and the present status of Japanese eel resources. We also describe the history of eel fisheries and aquaculture in Japan, and document recent progress in research on artificial production of glass eels, which may help eel conservation by reducing the impact of fishing glass eels in the wild. There is also reference to historical aspects of the utilization of eels, the distribution of eel products, and the role of eels in Japanese food culture. Finally, the cultural aspects of eels in arts and crafts are explored, and some traditions and spiritual relationships between eels and humans in Japan are described.

The Science of Eels in Japan

Taxonomy

Scientific studies on the taxonomy of Japanese freshwater eels started in a Dutch trade factory at Dejima in Nagasaki. Dejima was a fan-shaped artificial island with a surface area equivalent to two football fields. Built in the innermost recesses of Nagasaki Bay, it was connected to the mainland by a small bridge across a canal 15 m wide. Under the Tokugawa shogunate, when Japan was closed to the outside world for more than 200 years (1639–1854), Dejima was a solitary window for trade with the Dutch, the only western nation allowed to trade with the country.

Medical doctors attending the Dutch trade office at Dejima included the German Engelbert Kaempfer (1651–1716), the Swede Carl Thunberg (1743–1828) and the German Philipp Franz von Siebold (1796–1866), together with his assistant apothecary Heinrich Bürger (1806–1858). All these men helped introduce knowledge of Japanese customs, culture, plants and animals to European countries, but it was Kaempfer who in 1727 mentioned Japanese freshwater eels for the first time in European literature. In his posthumous publication "The History of Japan" (Fig. 6.1), he wrote: "Unagi is the common eel. Ōunagi is another kind of eel much larger than the common one." However, the eel he depicted in his book was the giant mottled eel, *Anguilla marmorata*, not the most common eel in Japan, which is the Japanese eel *Anguilla japonica*.

Von Siebold was sent to Japan as a physician in the Dutch office at Dejima. He had a mission too to gather as much

M. Kuroki (✉)
The University Museum, The University of Tokyo, Hongo, Bunkyo, Tokyo 113-0033, Japan
e-mail: mari@um.u-tokyo.ac.jp

M.J.P. van Oijen
Naturalis Biodiversity Center, 9517 Leiden, The Netherlands
e-mail: Martien.vanOijen@naturalis.nl

K. Tsukamoto
College of Bioresource Sciences, Nihon University, Fujisawa, Kanagawa, 252-0880, Japan
e-mail: tsukamoto.katsumi@nihon-u.ac.jp

K. Tsukamoto and M. Kuroki (eds.), *Eels and Humans*, Humanity and the Sea,
DOI 10.1007/978-4-431-54529-3_6, © Springer Japan 2014

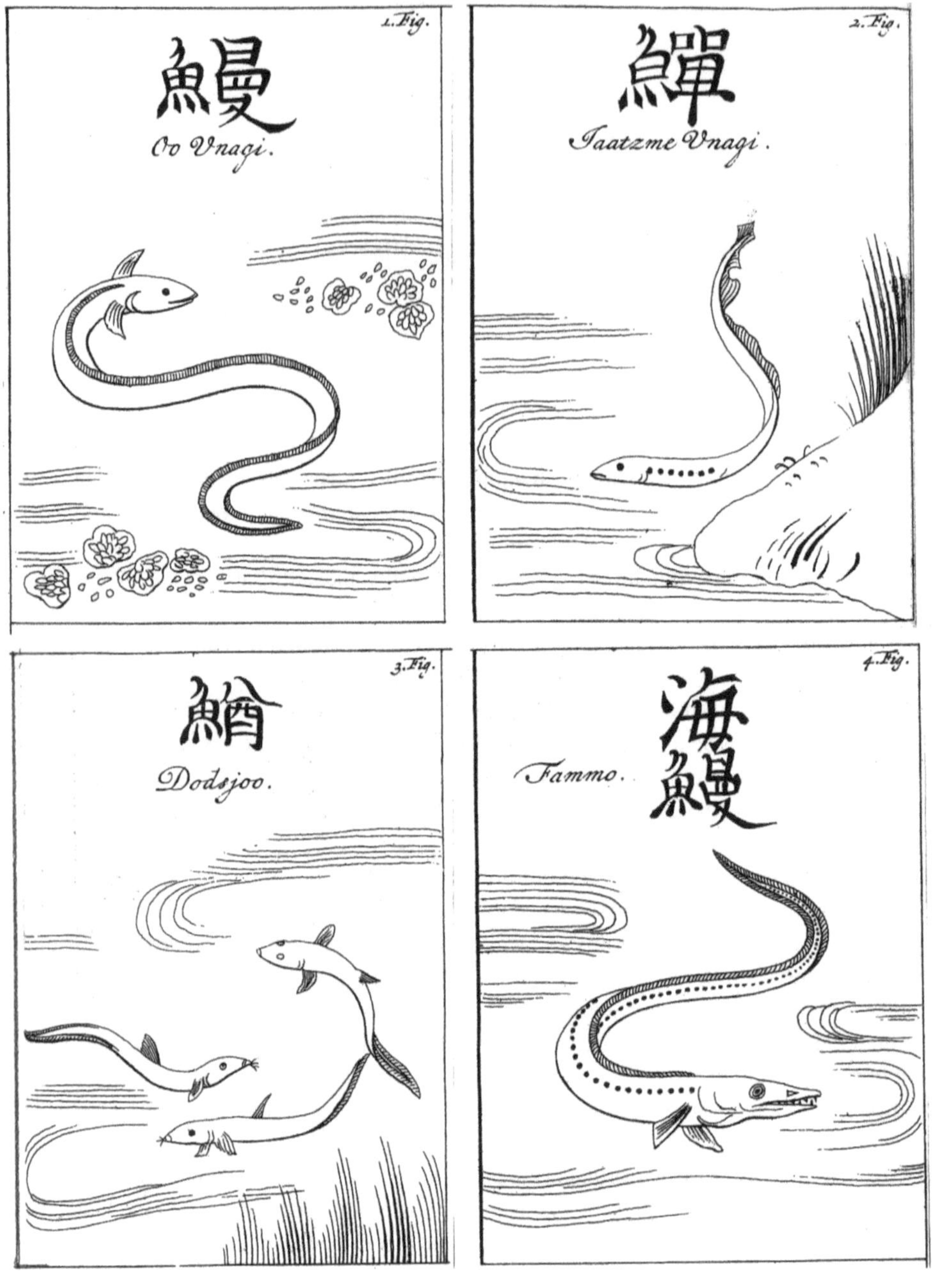

Fig. 6.1 Four eel-like fish depicted in "The History of Japan" by Engelbert Kaempfer (1727). *Top left*, giant mottled eel (freshwater eel); *top right*, lamprey; *bottom left*, loach; *bottom right*, pike eel (Source, RMNH, Netherlands)

information as possible on Japanese natural products, although the Dutch government knew, of course, that such undercover activity was officially prohibited. Nevertheless, he amassed huge botanical and zoological collections during his stay at Dejima from 12 August 1823 to 31 December 1829. Among ~700 fish he collected were four specimens of Japanese freshwater eel, all of which were preserved in alcohol and shipped first to Batavia (Jakarta, Indonesia) and then on to the Netherlands. After von Siebold left Japan, his assistant Bürger completed his mission and prepared a description of the freshwater eel (Fig. 6.2) along with descriptions of many other species, with accurate watercolour illustrations by the Japanese painter Kawahara Keiga in Nagasaki. Instructed by von Siebold, Bürger gave the species the provisional name *Muraena unagi*, a combination of the contemporary Latin genus and the Japanese name. The Japanese word *unagi* in his description refers to the freshwater eel.

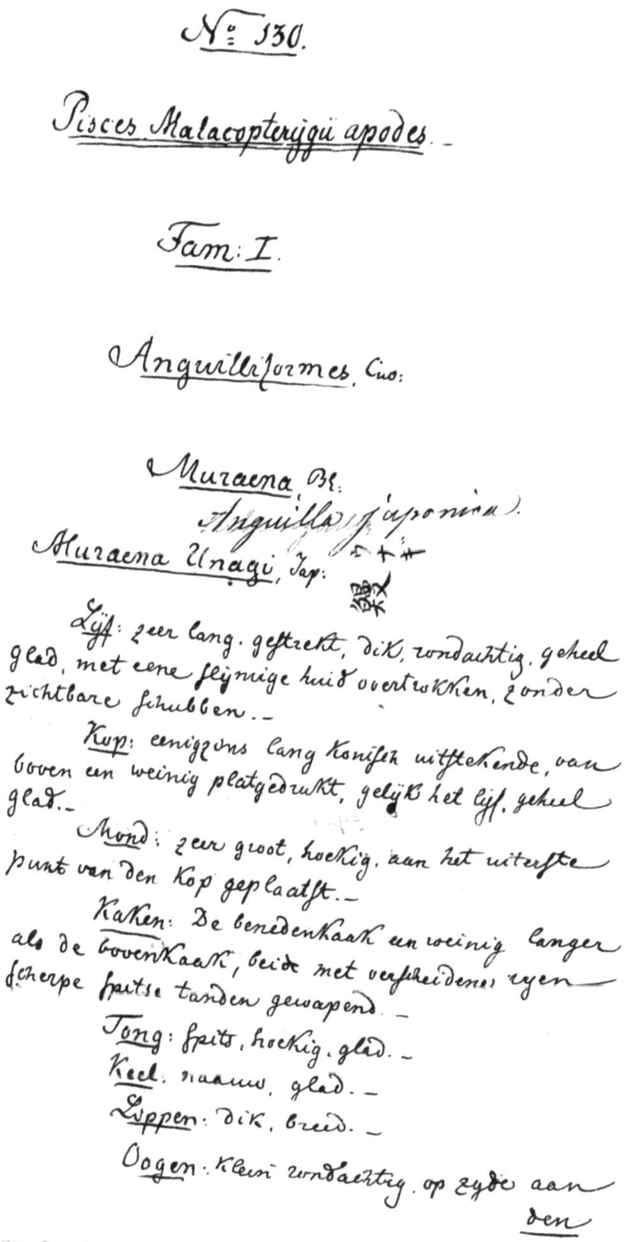
N°: 130.

Pisces Malacopterygii apodes. -

Fam: I.

Anguilliformes. Cuv:

Muraena. Bl.
Anguilla japonica.
Muraena Unagi, Jap: 鰻

Lijf: zeer lang, gestrekt, dik, rondachtig, geheel glad, met eene slijmige huid overtrokken, zonder zichtbare schubben. -
Kop: eenigzins lang konisch uitstekende, van boven een weinig platgedrukt, gelijk het lijf, geheel glad. -
Mond: zeer groot, hoekig, aan het uiterste punt van den Kop geplaatst. -
Kaken: De benedenkaak een weinig langer als de bovenkaak, beide met verscheidene rijen scherpe spitse tanden gewapend. -
Tong: spits, hoekig, glad. -
Keel: naauw, glad. -
Lippen: dik, breed. -
Oogen: klein rondachtig, op zijde aan den

Fig. 6.2 Heinrich Bürger's description of a Japanese eel tentatively called *Muraena unagi* (part of the front page). Source, RMNH, Netherlands

Based on Bürger's description, Keiga's watercolour illustration (Fig. 6.3) and the five specimens of Japanese eel sent by von Siebold (four) and his assistant Bürger (one) from Dejima to the Rijks Museum van Natuurlijke Historie (RMNH) in Leiden, Hermann Schlegel (1804–1884), curator of vertebrates at RMNH, described *A. japonica* in a section on Fauna Japonica together with Coenraad Jacob Temminck (1778–1858), the first director of RMNH. Thus it was that the Japanese eel from East Asia made its debut in the scientific community of the western world.

Spawning Area Surveys

Inspired by the discovery of the spawning area of Atlantic eels in the Sargasso Sea by Danish oceanographer Johannes Schmidt in 1922 (for details, see the Danish chapter in this book), Japanese researchers started surveying spawning areas around Japan in the 1930s. However, it was not until 1967, 30 years after the start of these surveys, that the first leptocephalus (the transparent leaf-like larva of an eel) was collected in the Bashi Strait near Taiwan. As the surveys proceeded farther upstream along two ocean currents, the Kuroshio and the North Equatorial Current, ever smaller leptocephali were collected from increasingly southern and eastern areas of each current. In that way, the surveys gradually approached the spawning area of the species, although progress was slow. Nevertheless, progress in our understanding of eel spawning ecology over the past decade has been remarkable; for example, genetically identified preleptocephali were successfully collected in 2005 (Tsukamoto 2006), spawning adults in 2008 (Chow et al. 2009; Kurogi et al. 2011) and even eggs in the estimated spawning area in 2009 (Tsukamoto et al. 2011). Therefore, it is possible now to predict with accuracy the spawning site of Japanese eels, based on oceanographic conditions and geological features. In the near future, it may be possible too to see spawning aggregations of adult eels under the darkness of the new moon near the seamount chain of the West Mariana Ridge.

Other Fields of Research

A major contribution made to the knowledge of eel biology has also been the discovery of Japanese and European (*Anguilla anguilla*) eels that never migrate into freshwater, spending most of their lives in estuarine or coastal areas: "sea eels" (Tsukamoto et al. 1998). That discovery overturned the conventional understanding of catadromy among freshwater eels, and had a ripple effect on strategies for resource management and conservation of eels in general. The pattern of facultative migration by sea eels is interpreted as some kind of atavism of a species that originated in the sea, and is found in many eel species. In fact, strong evidence for the oceanic origin of freshwater eels was obtained recently, based on phylogenetic analysis of whole mitochondrial genome sequences from 56 species representing all 19 anguilliform families. The results unequivocally show that freshwater eels originate in midwaters of the deep ocean (Inoue et al. 2010). Moreover, a molecular phylogenetic study of all 18 species and subspecies of *Anguilla* known revealed that *Anguilla borneensis*, a species endemic to Borneo, was the most ancestral form. That species originated near Borneo around the late Cretaceous, 60 Ma, and dispersed to the Atlantic Ocean through the ancient Tethys Sea

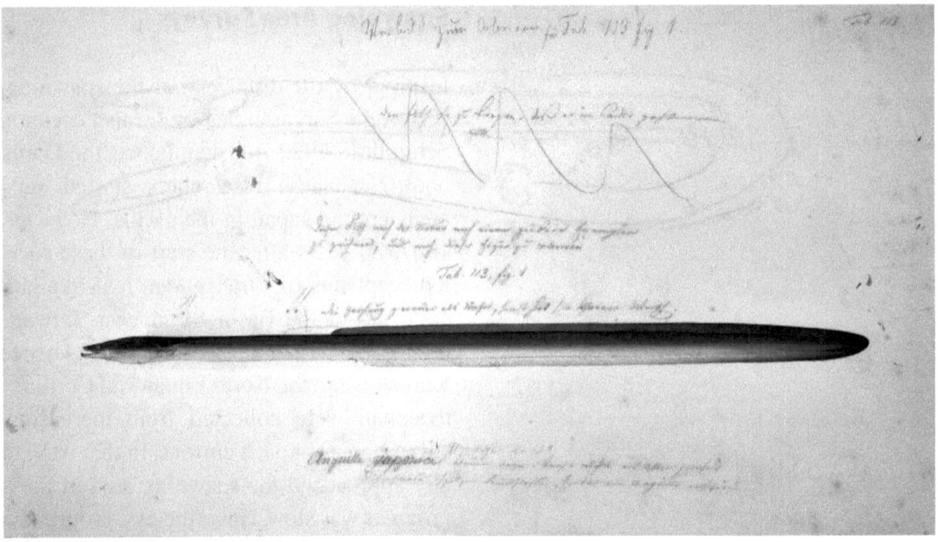

Fig. 6.3 Kawahara Keiga's watercolour of a freshwater eel with Schlegel's remarks and instructions in pencil for the lithographer (the specimen depicted has a short dorsal fin and might not be a Japanese eel, but possibly *Anguilla bicolor*, which could have been transported from Batavia, Jakarta, Indonesia, to Dejima in Japan. Source, RMNH, Netherlands)

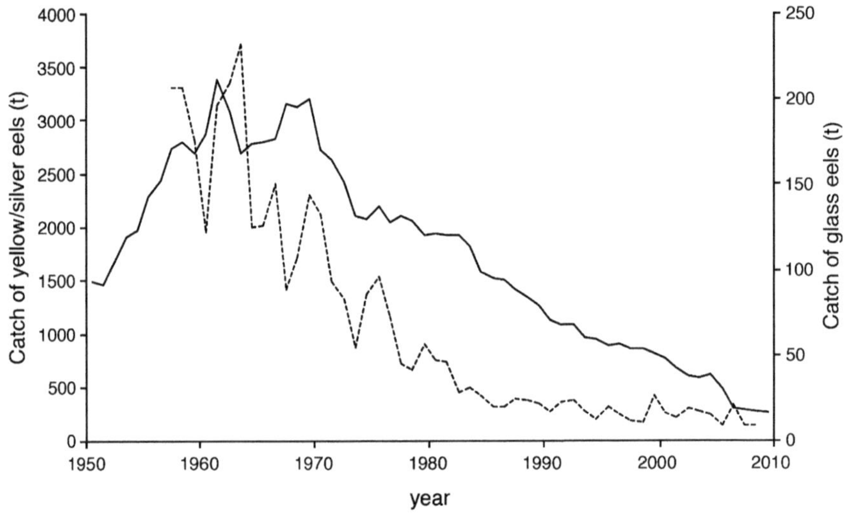

Fig. 6.4 Catches of glass eels (*dotted line*) and yellow/silver eels (*solid line*) in Japan

(Aoyama et al. 2001). Ecological aspects of tropical eels were also actively studied in terms of early life history (Kuroki et al. 2006), glass eel recruitment (Sugeha et al. 2001) and downstream migration of silver eels (Hagihara et al. 2012). The recent discovery of a new species, *Anguilla luzonensis*, from Luzon Island, the Philippines, should also be mentioned, because it has been a long time (about 70 years) since the Danish taxonomist Vilhelm Ege (1887–1962) established his stunning system of the genus *Anguilla* (Watanabe et al. 2009; Kuroki et al. 2012).

Resources and Conservation

In Japan, catches of glass eels in estuaries and of eels in more advanced stages (yellow and silver forms) in rivers have been decreasing over the past few decades (Fig. 6.4). Recent catches of glass eels in Japan are indeed only ~5 % of their peak levels in the 1960s, and catches of yellow and silver eels in rivers are also a mere fraction of what they were in the 1960s.

As mentioned several times in this book, possible causes of the decline in eel populations have been identified as overfishing, the degradation of riverine environments, regime shifts in the ocean, mortality in the turbines of hydroelectric power plants, other dams, parasites and pollutants. The first three factors in particular appear to be significant in the case of the Japanese eel. All of the factors, however, derive from human activities; even oceanic regime shifts may be linked to such anthropogenic factors as *El Niño*, climate change and global warming (Kim et al. 2007).

Overfishing is apparently a major factor in the long-term decline of eel resources. Some fishers taking glass eels claim to have caught at least 70 % of the glass eels recruiting into the river mouths where they were fishing for them (H. Omori, personal communication). Construction channelling structures and riverbank repairs have reduced habitats for eels and their prey by artificially adding concrete structures to natural waterways. Water pollution in rivers, especially at river mouths (the most important habitat for eels), may also damage the gills and chloride cells that enable eels to osmoregulate in the frequently changing salinity environment of the river mouth. Catches of glass eels tend to be low in *El Niño* years, when there is a southward shift in latitude of the salinity front, affecting the latitude of spawning sites of adult eels in their spawning area (Kimura et al. 2001; Kimura and Tsukamoto 2006). A southward shift of just 1° in spawning site may cause the entrainment of eel larvae into the south-flowing Mindanao Current, reducing the percentage of larvae able to transfer from the North Equatorial Current to the Kuroshio, which current normally carries the larvae to their growth habitats in East Asia (Kim et al. 2007; Zenimoto et al. 2009).

Fluctuations in eel resources need to be considered by distinguishing between short-term (annual) and long-term (decadal, centennial or longer) phenomena. Overfishing of glass eels and silver eels and the degradation of river environments undoubtedly cause long-term decreases in eel stocks by reducing the number of silver eels migrating successfully back to sea. Climate change and global warming may also have a long-term effect on the resource, whereas short-term fluctuations in eel resources can be caused by monthly and annual changes in oceanographic conditions such as *El Niño* events, changes in seawater temperature and changes in the latitude of the salinity front at the spawning area and the bifurcation zone of the North Equatorial Current off Luzon Island. All these factors influence the recruitment of glass eels to East Asia in any year, so the fluctuations in eel resources we see today are almost certainly a mixture of these short- and long-term changes.

The most effective action for conserving the Japanese eel would be to increase the number of silver eels that start their spawning migration from the coastal waters of East Asia, by reducing fishing pressure on both glass eels recruiting to coastal waters and silver eels returning to sea. However, resource management of eel fisheries over their whole geographic range is not easy, because the Japanese eel is an international fish resource shared by Taiwan, China, Korea and Japan. It is clear that the administrative powers in each country have to organize an international committee for regulating eel fisheries fairly among East Asian countries if really effective management is to be put into place.

There has been one key development, however. To facilitate the conservation of Japanese eels as a common property, the East Asia Eel resource Consortium (EASEC, a non-governmental, non-profit organization) was established by scientists working on eels and eel industry representatives from Taiwan, China, Korea and Japan in 1998. Since then, a conference has been held annually. The "Eel River Project" has also been started on several selected rivers in each country, with the aim of monitoring the recruitment of glass eels all year round and making those rivers eel sanctuaries or model rivers in the near future. Such activity will hopefully stimulate public interest in eel conservation in local society and may even foster the clean-up of estuarine areas by improving water quality and suppressing the discharge of pollutants by each country.

Fisheries

In general, eels of any species are not easy to catch. Various fishing gears and methods have been developed in each country that has eels for catching each growth stage and at different sites, such as beaches, rocky banks, muddy ponds and the lower or upper reaches of rivers (see the other chapters in this book too).

Glass Eel Fisheries

At the glass eel stage in their development, eels are small and feed little, so simple and easy techniques are used to catch them as they recruit into estuaries; these include small handnets (*tamo-ami* or *tama-ami*), relatively large scoopnets and small setnets (*mekko-ami*), using the tidal flow in the estuaries. A set consisting of a handnet and lamp is typical fishing gear at night around the new moon, and the familiar sight of numerous lamps in the darkness of cold winter nights in Japan brings with it the real feeling of winter.

Yellow Eel Fisheries

Eels at this stage feed actively on benthic animals in coastal waters, estuaries, rivers, ponds and lakes at night. During daylight, they tend to hide in refuges among rocks, in burrows of mud, or under fallen timber on the bottom of aquatic habitats. Using this behavioural knowledge, many different types of fishing gear and methods have been developed to catch them. Hook and line (*okibari*), longline (*haenawa*), bamboo sticks with hooks at the tip and earthworms inserted into holes in rocks (*anazuri*) are all fishing methods used in Japan to attract eels with bait. One local traditional fishing method called *zuzuryo* (or *suzuryo*) uses a long wooden pole with many earthworms in rings at the tip to catch eels that bite the bait. Further, eel pots made of bamboo netting (*uge* or *tsutsu*, normally about 70 cm long and 10 cm in diameter) are a typical method of catching eels by attracting them with bait and preventing their escape with a trap at the entrance.

Sections of large hollowed-out bamboo or simple plastic pipes (*unagi-zutsu* or *suppon*) are used as refuges for eels

Fig. 6.5 Traditional methods of fishing for eels in Japan. (*left*) *Ishiguro-ryo* rock-pile eel fishing (from *Mie-ken Suisan Zukai*, published in 1883; source, Mie). (right) *Unagi-kaki* fishing with an *unagi-kama* big eel scythe (from the *ukiyo-e "Tōto Miyatogawa no Zu"* by Utaga Kuniyoshi, published in 1830–1844; source, Ōshima-ya Collection)

attracted to their offer of shelter. Without traps or bait, these are attached to long ropes and laid underwater on the bed of the waterbody, allowing eels to enter and hide, and to be harvested periodically. Another use of refuges is the *shibazuke*, a big bundle of twigs (*bosa*) that provides shelter for small eels, which are then scooped up with the bundle into a large handnet, catching the eels inside. Yet another traditional method of eel capture in Japan is the *ishiguro-ryo* or *unagi-gura*, mounds of rocks built at low tide (Fig. 6.5) in shallow estuaries. At a subsequent low tide, fishers enclose the mound with a net or a bamboo or reed screen, then remove the rocks to catch the eels hiding inside the mound, using eel scissors to grasp the eels, or driving them into an eel pot attached to the screen. This unique technique of *ishiguro-ryo* has traditionally been used on the Japanese islands of Kyushu and Shikoku.

Other fishing methods for eels are *unagi-zuki*, using a spear to catch eels visible in water, and *unagi-kaki*, using *unagi-kama* eel scythes to extract the eels hiding in mud by searching at random with a scythe (Fig. 6.5). Another large-scale fishing method is *kaibori*, in which people drain a pond or a part of a river to trap stranded eels. Finally, small-scale setnets (*machi-ami*, *kakudate-ami* or *fukuro-ami*) are sometimes used to fish for eels in rivers, lakes and coastal waters.

Silver Eel Fisheries

Eels at the silver stage stop feeding and their behaviour changes. Whereas yellow eels stay at the bottom and hide in refuges during daylight, metamorphosed silver eels become active even during daylight and swim in the upper layers of waterbodies. The changes appear to be in preparation for a long spawning migration across the ocean. As silver eels do not eat so cannot be attracted by bait or refuge, they can only be caught by *yana* (weirs) or *machi-ami* (setnets), or by *dōjiri* (big eel pots) during their downstream movement in rivers or offshore migration in coastal waters, or by random chance operation of eel scythes in muddy estuaries. For a limited period in winter, people may be able to scoop silver eels with handnets from the surface of coastal waters at night, though that method is not that well established as a means of fishing eels.

Aquaculture

With Japanese consuming as much as 100,000 t of eels annually, aquaculture is an important industry (Ida 2007). Indeed most (>99.5 %) of the eels consumed these days are cultured, having been reared from wild-caught glass eels in aquaculture ponds. In Japan, glass eels of body weight 0.2 g are reared to about 200 g in just half a year, an increase in body weight of 1,000 times at a growth rate of about 1 g day^{-1}. However, variation in the growth rate of eels is massive, so the eels are regularly sorted for size during the rearing period. Big eels of >500 g are often called *boku*, showing overgrowth to commercial size in culture ponds. It is known that most cultured eels become males and that cultured females are rare, although the reason for this is unclear, despite speculation that the high temperatures and density of fish in ponds may cause a bias towards sex differentiation into males.

Kurajiro Hattori (1853–1920) started eel aquaculture in Japan using a 2 ha pond at Fukagawa, Tokyo, in 1879. He used elvers, not glass eels, as seedlings; glass eels were not used for aquaculture until the 1920s. Initially, eel aquaculture was popular in the Tokai area in the centre of Honshu Island, using outdoor ponds (Fig. 6.6). The practice then

Fig. 6.6 An aerial photo of eel ponds in Shizuoka, Japan, in 1969 (photo, Yoshida, Shizuoka)

spread to the Mikawa area several decades later. Eel aquaculture yielded 3,000 t or more in the 1930s and soon exceeded the catch of wild eels. Eel aquaculture temporarily declined in production during the Second World War, but it recovered in the 1960s to produce more eels than it did before the war. In 1955, waterwheels were introduced to ensure a sufficient supply of oxygen into the ponds. The development of assorted feed for eel aquaculture started in 1964; before that, eels had been fed with cheap boiled sardine or sandlance. Assorted feed enabled eel farmers to enhance growth and to ensure efficient, scheduled production. In 1972, intensive aquaculture in vinyl houses using heated water became the mainstream method, overtaking the previous extensive use of single outdoor ponds. This intensive means of eel aquaculture extended to Shikoku and Kyushu in western Japan in the 1970s (Kuroki and Tsukamoto 2012).

Japan's production of cultured eels peaked at 40,000 t in 1989 (Ida 2007). In Taiwan, full-scale eel aquaculture was started in the 1960s, gathered speed with the onset of severe fish diseases in Japan in 1968, and finally overtook eel production in Japan by around 1990. However, the 1990s saw a rise in Chinese eel aquaculture as it took advantage of low labour costs and inexpensive feeds, and China's annual eel production quickly exceeded 40,000 t, more than both Japan and Taiwan. That success was also based on the aquaculture of European glass eels, which Chinese traders imported in large numbers from Europe, rearing them inland at low water temperature but taking a longer time (the so-called "running water method"). In 2000, however, a prohibited antibacterial drug was detected in Chinese eel products on the market and all eels exported from China became subject to strict regulation. This, together with the increasing shortage of seedlings for aquaculture from Europe as a result of declining recruitment there and new CITES regulations to protect the European eel, caused a decline in China's eel aquaculture from the turn of the millennium. Therefore, over the years, eel aquaculture in East Asia has been disrupted and changed by waves of price fluctuations and a range of fish diseases, varying catches of glass eels over time, and strict regulation on the movement of product.

Artificial Production of Glass Eels

Seedlings for eel aquaculture have been totally dependent on wild glass eels or elvers caught by fisheries. However, the recruitment of glass eels fluctuates greatly from year to year, and resources have been declining dramatically for many years. Therefore, to obtain a stable supply of seedlings for aquaculture, research on artificial glass eel production has been underway in Japan for 50 years.

It is difficult to enhance the maturation of adult freshwater eels without intense artificial manipulation, such as using exogenous hormone treatment, temperature controls or nutritional enrichment. In comparison, red sea bream, Japanese flounder and even pike eels can be matured merely by raising the temperature slightly or by providing the fish with one or two injections of hormone. Larvae are also difficult to rear to the glass eel stage; even wild leptocephali take as long as 3–6 months to reach that stage after hatching, and reared ones take 1 or 2 years to become glass eels because of their extremely slow rate of growth. An ideal diet for artificially reared eel larvae has yet to be found, despite many years of effort being put into finding one. Putting the above into perspective, freshwater eels may well be the most difficult species that humankind has ever attempted to spawn and rear using aquaculture techniques.

At the dawn of this type of research, French physiologist Maurice Fontaine was the first to succeed in enhancing spermatogenesis in male European eels by using the urine of a pregnant woman (Fontaine 1936). In 1961, Japanese scientists Hideo Sato and Takashi Hibiya of the University of Tokyo succeeded in artificially evacuating sperm from a Japanese eel by stimulating it with human chorionic

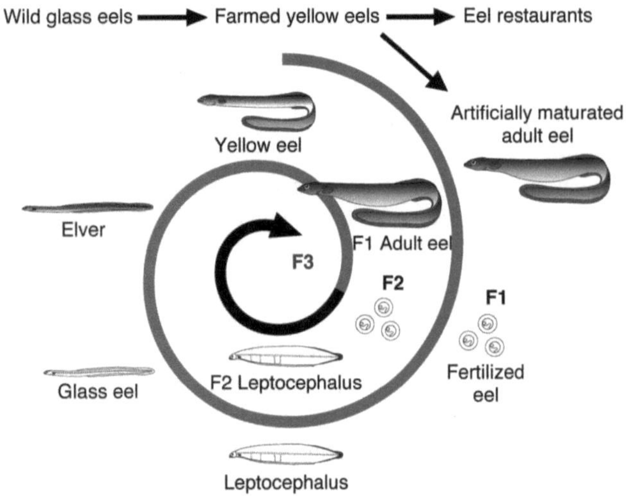

Fig. 6.7 The cycle of artificial mass production of glass eels to produce aquaculture seedlings with the aim of establishing a domestic strain of freshwater eel in Japan

gonadotropin (Satoh et al. 1962). Then, in 1973, Kiichi Yamamoto, Kohei Yamauchi and their colleagues from Hokkaido University managed to hatch larvae of the Japanese eel from artificially fertilized eel eggs by injecting them with salmon pituitary extract (Yamamoto and Yamauchi 1974), and succeeded in rearing the larvae for another 14 days (Yamauchi et al. 1976). All these were pioneering achievements in the challenge of artificial glass eel production. In 2001, Hideki Tanaka and his colleagues of the National Institute of Aquaculture, Fisheries Research Agency of Japan (FRA), succeeded in obtaining the first glass eel artificially produced in the laboratory by feeding eel larvae a diet mainly of shark egg yolk (Tanaka et al. 2003). Then, in 2010, the FRA successfully obtained the larvae of a second-generation (F2) of artificially produced Japanese eels by rearing laboratory-produced glass eels to adulthood for egg production, completing the loop of a whole lifecycle of freshwater eels under human control (Fig. 6.7). Now that F2 generation is ready to produce F3 larvae. Thus far, however, the Japanese eel is the only species of eel in the world to have produced glass eels under controlled, non-wild conditions; the larvae of other eel species have so far only survived through the preleptocephalus stage, or a little longer.

Despite the advances made so far, however, a technique for the mass production of artificial glass eels for aquaculture has still to be developed, notwithstanding the more than 50 years of dedicated effort expended to this end. According to a rough estimate based on Asian market needs, hundreds of millions of glass eels for aquaculture will be required every year, but to date we can produce only several hundred annually in the laboratory. This suggests that a further breakthrough in research is needed for practical mass production of artificial glass eels to be achieved. Success in such a challenge would reduce the obviously decimating human impact on wild eel stocks, and perhaps help endangered eel species to a better conservation status in future. Moreover, we will need to breed strains appropriate for aquaculture, with good taste, excellent survival rates, short larval duration, disease tolerance, stress resistance, etc, although these artificially produced eels will then have to be kept strictly as domestic strains and not be released into the wild, to preclude genetic disturbance and the possible spread of disease.

Archaeological Sites with Eel Bones

When and where on earth did humans start to use eels? The origin of the genus *Anguilla* is estimated to date back about 60 million years, whereas early humans only appeared a million years ago, at most. This means that eels were already in rivers when humanoids appeared on earth. Eels were a likely top predator in freshwater habitats, with few notable natural enemies, and for that reason, they appear to have behaved and still behave in their own ways in the habitats they occupied and still occupy in much of the world. This can be imagined easily when we see the abundance of carefree, languid eels in Tahiti in Polynesia or Pohnpei in Micronesia, where eating eels is taboo and people hold them in respect. In such situations, ancestral human beings must have been able to catch them easily by hand without the need for sophisticated fishing gear. Eels in shallow waters near the colonies of our ancestors must have been an important food, because eels are nutritious, rich in fat and vitamins, and likely one of the best food resources for surviving the rigours of winter.

Eel bones (vertebrae) have been found at 120 sites from the Jomon period (2,500–16,500 years ago) and the Yayoi period (1,750–2,500 years ago) in Japan (Fig. 6.8; Morse 1879). Eel bones found in shell mounds obviously show that people in the Jomon and Yayoi periods used eels frequently as food. As eels recruit to Japan using the Kuroshio Current, a western boundary current in the North Pacific, most of the sites with eel bones are located along the Pacific coast of the Japanese archipelago along the Kuroshio. In particular, there are many such sites along the coasts of Tokyo Bay and Sendai Bay, but just two that face the Japan Sea or the East China Sea. The northernmost of them all is the Misawa-4 site in Hokkaido, suggesting that a much warmer climate than at present once permitted eels to inhabit that island. The southernmost is the Chibatsukahara site in Okinawa, one of the southern islands of Japan. Considering the present distribution of *Anguilla* in that area, the bones there could possibly be those of the giant mottled eel, *A. marmorata*.

Eel bones have also been found at the site of the former samurai residence of the Kaga clan (Han) in the Edo era, located on the University of Tokyo's Hongo Campus.

Fig. 6.8 The Omori shell mounds being excavated (from *Shell Mounds of Omori* by Edward Sylvester Morse 1879). *Insert*, eel vertebrae from the middle of the late Jōmon period (~4,000 years ago), excavated from the Miwanoyama shell mound in Nagareyama, Chiba, Japan

Judging from the caudal vertebra where the neural arch and vertebral body attach to each other, these bones are from *A. japonica*, not members of the Congridae or the Muraenidae. There were also many finds of ceramics, pottery, oriental pipes, sake bottles and *go* pieces at the site. That part of the residence was for samurai who took up posts in Edo, the capital at that time, leaving their families in far-off Kaga. To console themselves in their loneliness, they appear to have purchased live eels in downtown Edo, then grilled them in the backyard of the residence to eat, washing the meal down with sake.

Distribution of Eel Products

As already stated, Japan currently consumes about 70 % of the eels produced worldwide by aquaculture. As the primary flow of eels around the world in modern times, glass eels of the European eel *A. anguilla* caught in the wild were until recently brought to China to be cultured, and the eventual products then exported to Japan. The demand for glass eels as aquaculture seedlings is always high, and the price of Japanese glass eels per gramme sometimes seems to exceed that of gold. However, the lack of wild-caught Japanese eels induced a rapid increase in the price of foreign glass eels, and at one time the harvest and distribution system became chaotic in some parts of the world. In 1990, just 18 t of glass eels were transported from Europe to Asia, but this figure increased dramatically to as much as 485 t in 1997, when the catch in Japan was a mere 20 t (Ida 2007). Poaching was rampant and there even appeared to be systematic poaching by illegal groups armed with weapons; many of these poached glass eels were transported by air from Madrid to China. Guns were drawn at times too in the northeastern US, in what was called a "gold rush" for glass eels of the American species *Anguilla rostrata* to be sold to Asia. In Asia too, poaching and smuggling were rampant, the sudden rise in glass eel prices causing turmoil in the fisheries and in some places prompting new regulations to be enacted.

Eels are, in general, distributed as live fish. *Domaru* or *dōman* baskets are special transport containers for eels that are round (ca. 50–70 cm in diameter ×30 cm high) and made traditionally of woven bamboo. Today, however, they are generally made of black-meshed plastic. These *dōmaru* containers have a round concave lid of the same material, and many *domaru* with eels inside can be stacked into a column. A small amount of running water can be supplied to the top of the column, from where it drips down though the column and keeps the contained eels wet. The places where the *dōmaru* are stacked up are called *tateba*, which means a place (*ba*) for piling up (*tate*) the *dōmaru*. A *tateba* is a transit station or pick-up point for live eels. The *dōmaru* eel containers and *tateba* eel-holding method constitute a specially developed distribution system for live eels, making use of eels' ability to survive in moist conditions, out of water, for some time.

In the middle of the Edo era, there was a special route for eel transportation called the *unagi-kaidō* (eel road) in the Chugoku region of western Japan (Kuroki and Tsukamoto 2012). This was used to transport many of the eels caught in large quantities in the Nakaumi lagoon in Izumo, on the coast of the Japan Sea. The peak period for this "eel road"

Fig. 6.9 The "Izumo-ya" eel restaurant founded in 1876 in Osaka. *Top:* A commemorative photograph taken in front of the restaurant near the end of the Meiji era (1912). The signs outside the restaurant declare "From Izumo, the best in Kansai" and "Izumo-ya Unagi Mamushi" (steamed or mixed eel rice). *Bottom:* A truck from the US Ford company used by Izumo-ya to transport eels during the early Showa era (mid-1920s). On its side is a large advertisement for "Izumo-ya" in Japanese (photo, The Sanyo Shinbun)

was from the end of the Edo era to the early Meiji era (around 1870), when 56 t of eels per year were transported from Izumo to Osaka, a big city then, with a huge market. This large-scale transportation of live eels had a huge impact on food culture in Osaka, about 300 eel restaurants named "Izumo-ya" (*ya* means a shop or house) being clustered close together in one part of the city (Fig. 6.9). In those days, "Izumo-ya" was synonymous with eel restaurants.

Sajū Matsumoto, a merchant from the Matsue Clan, had the idea to transport live eels collected from Izumo to Osaka in the Edo era. He obtained permission from the Matsue Clan and purchased the authorization of Shōgo-in Temple in Kyoto, a powerful authority at that time, for the safe and smooth transportation of eels between Izumo and Osaka. Shōgo-in Temple gave him a flag and a lantern (*hako-chōchin*) as a sign of approval. A big team of 20–30 strong young men was organized, each carrying 20–30 kg bundles of eel containers on their shoulders. First they crossed the mountains, with some stops on the way to keep the eels wet, before arriving at Katsuyama in Okayama. From there they took a *takase-bune* river boat and went down the Asahi River to the port of Okayama, where they transferred from the river boat to a specially designed seagoing ship with a well to hold live eels, which sailed through the Seto Inland Sea to Osaka. The whole trip from Izumo to Osaka, where the live eels were sold at a good price, took about 7 days.

Eel Cuisine

Together with the two well-known Japanese dishes *sushi* and *tempura*, eel *kabayaki* is one of Japan's representative foods (Kuroki and Tsukamoto 2012). *Kabayaki* is grilled eel with soy sauce, often served with rice in eel dishes such as *unadon* (a rice bowl with *kabayaki* on the top) or *unajū* (rice in a lunch box with *kabayaki* on top; Fig. 6.10a). *Una* is an abbreviation of *unagi* (eel), *don* comes from *donburi* (bowl), and *jū* is short for *jūbako* (piled lunch boxes). There are many other eel dishes in Japan, including *shirayaki* (simple grilled eel without soy sauce; Fig. 6.10b), *hitsumabushi* (a mixture of chopped *kabayaki* and rice; Fig. 6.10c), *seiromushi* (steamed *kabayaki*), *uzaku* (chopped *kabayaki* with pickled vegetables; Fig. 6.10d), *umaki* (rolled eel omelette; Fig. 6.10e), *kimosui* (soup with eel liver; Fig. 6.10f), and *unagi-zōsui* (rice porridge with chopped *kabayaki*).

The name *kabayaki* appears in *Suzuka-ke ki* ("Documents of the Suzukas," 1399), a family record of the main Shinto priests at Yoshida Shrine in Kyoto. The origin of the name comes from the old style of cooking eels in those days. Chopped eel would be grilled on a skewer and served with *miso* (fermented soybean paste) and Japanese pepper. As its shape resembled the cattail (*gama*), the cooking style was called *gamayaki* (*yaki* means "grill"), which was later changed to *kabayaki* (in the essay *Katabisashi* written by Hikomaro Saitō in 1853). Later, people started to make fillets of eel and grilled them with soy sauce, which was also slightly sweetened, so establishing the present-day favourite *kabayaki*. Grilling eel fillets is unique as a cooking method and is only common in Japan and in Japanese eel restaurants in Paris, Seoul and Taipei. This is the most significant characteristic that makes *kabayaki* very popular in Japan and a globally unique eel dish.

An illustration of a street stall selling *kabayaki* can be found in the popular book *Kōshoku ubuge* written by Horie Rinkō in the Edo era. It shows fillet-style *kabayaki* and the words *unagi sakiuri*, meaning "eel fillet." That work shows that the custom of cooking eel fillets in Japan started at least 300 years ago. Originally, there was a tradition of eating pike

Fig. 6.10 Traditional eel cuisine in Japan: (**a**) *kabayaki* (grilled with sweetened soy sauce); (**b**) *shirayaki* (grilled without sauce); (**c**) *hitsumabushi* (minced grilled *kabayaki* mixed with rice, popular in Nagoya); (**d**) *uzaku* (grilled and chopped eel with sliced cucumber in vinegar, sometimes with an added yam slice on top); (**e**) *umaki* (rolled omelette with grilled eel); (**f**) *kimosui* (eel liver soup). Photos **b**, **d**, **e** and **f**, Unagi Hyakusen

eels in Kyoto in summer, and these were cooked as fillets too. Therefore, it can be speculated that the trend of cooking eel fillets started in Kyoto or Osaka then spread east to Edo (Tokyo).

It is revealing to compare the culture and habits of the "Kanto" region of eastern Japan, represented by Tokyo, and the "Kansai" region of western Japan, with its two centres of Osaka and Kyoto (Kuroki and Tsukamoto 2012). *Kabayaki* differs between the two regions, because in Kanto the eel is

Fig. 6.11 Four representative types of knife used to prepare eels in Japan. From *left to right*, the Osaka type, a simple pointed blade, the Kyoto type, which is heavy and specially designed for splitting eels, the Nagoya type, a smallish kitchen knife, and the large Kanto (Tokyo) type, used to cut up eels as well as to split them. The Osaka type, originally used for preparing daggertooth pike conger, was probably the prototype, and the Kanto type has become differentiated into the most evolved form and function. Although used for the same basic purpose, these types of knife all remain in use today, reflecting local methods of preparation and the preferences of local eel chefs

slit from the dorsal side to prepare the fillet, whereas in Kansai, the eel is slit from the ventral side. There is some speculation as to why Kanto people do not use the ventral method. It is said that Kanto people associate it with the image of *harakiri* or *seppuku* (ritual suicide performed by samurai), because Kanto is the area where the samurai originated and the Tokugawa shogunate was located in Edo (Tokyo), the main city of Kanto. Moreover, another reason may be that *kabayaki* in Kanto has an additional cooking process of steaming to reduce the excess fat, whereas in Kansai the *kabayaki* is cooked directly without steaming. During grilling, the steamed ventral parts are easily broken if the eel is slit from the ventral side and the ventral parts located at the outer edge of the fillet.

There is also variation in the eel knives used in Japan, not only between Edo and Osaka, but also in other areas such as Kyoto and Nagoya (Fig. 6.11). The Osaka type is the simplest, whereas the Kanto type is big and has two functions: to cut the whole eel open with the tip of the knife and to carve it into fillets with the middle part of the blade. Much variation is evident in eel knives and how they are used, according to the locality and the workers processing the eels. The variation appears to have some relation to the grilling method and cooking style. For example, the Nagoya-style knife looks ordinary with no pointed tip, perhaps because there the eels are served as a dish called *hitsumabushi*, with *kabayaki* chopped into small pieces. In this case, the function of cutting the *kabayaki* into pieces is more important for the knife. Another point of anthropological interest would be to consider the phylogenetic relationship among the eel knives, to discover which was the original and how they all evolved with the spread of eel dishes.

To establish *kabayaki* as a dish, sauce is indispensable. Traditionally, it was made of soy sauce and *mirin* (sweet sake), the mixture of two tastes (salty soy and sweet *mirin*) yielding the most favoured combination for Japanese. The aroma of this sauce when browning on high-quality charcoal stimulates the appetite and reminds Japanese people of eel *kabayaki*. Moreover, the mixed flavour of this sauce and the fat in the eels gives *kabayaki* an incomparable taste.

Although it has a history of some five centuries in Kansai, soy sauce started to be produced in Kanto from around the middle of the Edo era, 250 years ago. *Mirin* first appeared as a seasoning in the cookbook *Manpō ryōri himitsubako* (a secret box filled with 10,000 treasured cooking recipes) published in 1785. This suggests that the thick *kabayaki* sauce with its salty–sweet taste could have been made about two centuries ago, when the eel *kabayaki* we know today was finally established.

According to *Edo Kaimono Hitori Annai* (Nakagawa Hōzandō), an 1824 guide for travellers unfamiliar with Edo, there were 22 registered *kabayaki* restaurants in Edo at that time. The food and drink pages of the guide include a section on eel *kabayaki*, listing the names of many famous eel restaurants then. This fact alone shows that the culture of eating eels was already firmly rooted in the streets of old Edo and that these long-established restaurants must have vied with each other to hone their skills. In fact, a ranking chart called "Banzuke," including an evaluation of each eel restaurant, was published in 1852 and placed in libraries (Fig. 6.12).

For many people in Japan today, the term "Edo-mae" will conjure up images of *sushi* (Kuroki and Tsukamoto 2012), but in Edo days, the term was used in the sense of "fish and shellfish caught in the sea, rivers and canals in front of Edo Castle (*Edojō-mae*)"; the most representative of these fish were eels. Eel *kabayaki* was a popular dish with the people of Edo, and there were numerous eel restaurants around Edo Castle. Initially, however, eels were sold mainly at street corners or from street stalls, because they were a common food. Some years later, eel restaurants started providing rooms for customers to enjoy *kabayaki*, which gradually changed to a delicacy for special occasions. Ever since Edo times, then, it has been possible to build a thriving business on the single taxon "eels," demonstrating just how popular eels have been with Japanese people.

In Japan, there used to be various taboos or general beliefs of foods that "should not be eaten together," and some of these combinations of food items also included eels

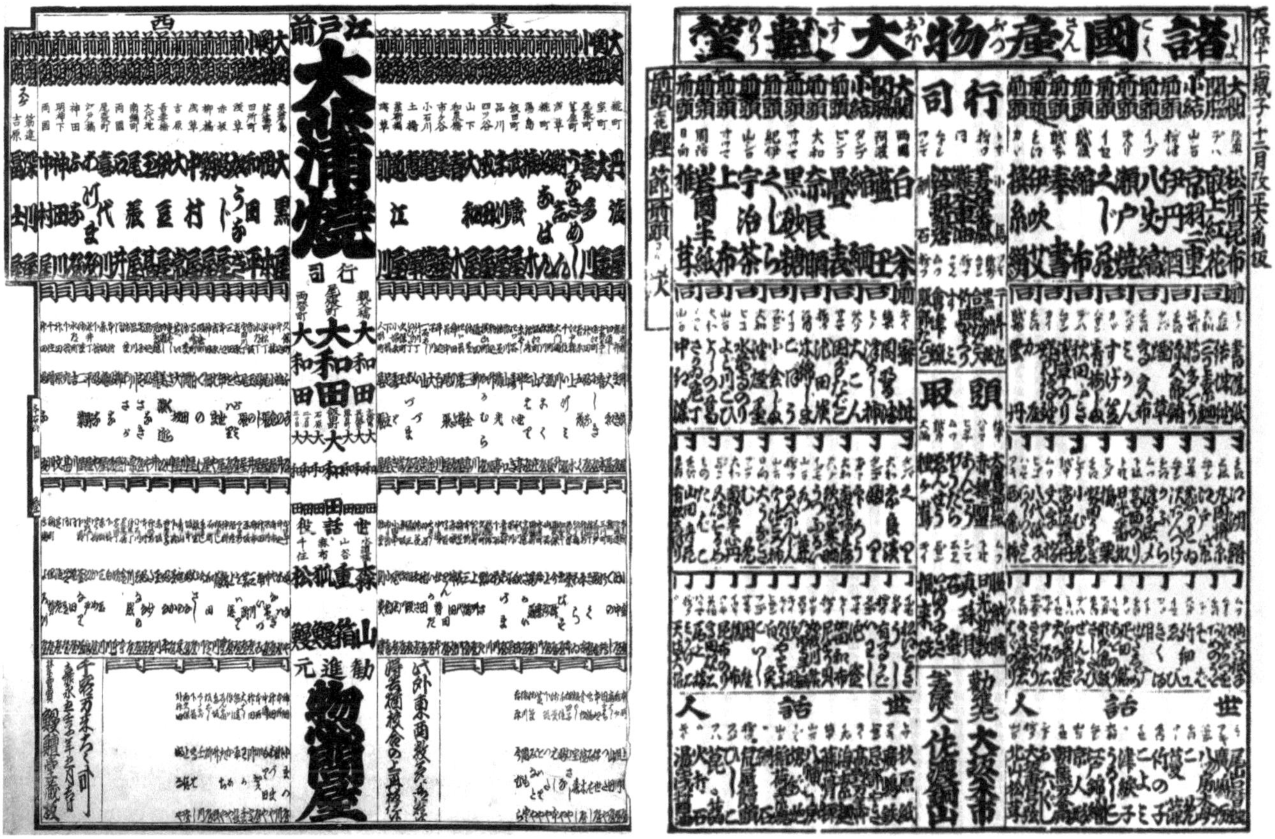

Fig. 6.12 Sumo-style ranking chart of 221 eel restaurants in Edo (*left*, *Edomae Ōkabayaki Banzuke*, 1852) and a ranking chart of special products from various districts in Japan with two famous eel localities, Seta and Hyuga (*right*, *Shokoku Sanbutsu Ōzumo*, 1840)

(Kuroki and Tsukamoto 2012). For example, eating *tempura* together with watermelon, or pond snails with *soba* noodles, could cause poisoning or diarrhoea. The "don't eat together" combinations for eels involve vinegar, ginkgo nuts and *umeboshi* (salted plum). Eels and *umeboshi* is the most famous of these taboo combinations, but there is no scientific evidence to suggest a problem with eating them together. In the early Showa period (around 1930–1960), there were many handbills showing different "don't eat together" combinations, distributed by door-to-door medicine salesmen from Toyama as free gifts for sales promotion. Housewives would put them up on their kitchen walls so that they could refer to the "don't eat together" combinations when planning their meals.

Words, Idioms and Poems

A horse can be called many things in Mongolian, depending on its sex, age and other features. There are also many words for "camel" in Arabic. The number of names given to a creature reveals the extent of its importance to humans. This is certainly true in the case of eels in Japan; in *Nihon Gyomei Shuran* ("List of Japanese Fish Names," 1958), Keizō Shibusawa lists 113 words for eel in Japanese, proof of the importance of eels to Japanese people.

Though normally known as *unagi* today, back in the Man'yō era (seventh–eighth centuries) eels were called *munagi*. Some say this derives from *mu-nagi* ("long body"), in reference to the elongate shape of an eel. Another theory is that *munagi* originally described eels in the yellow eel stage, when their "*muna* (chest)" is "*ki* (yellow)," *ki* being pronounced in Japanese as *gi*. Again, the early Edo Confucian scholar and botanist Kaibara Ekken (1630–1714) wrote, in his *Nihon Shakumyō* ("Japanese Etymology," 1700), that *munagi* was derived from *munagi*, the long round ridgepole that supports the roof of a traditional Japanese house. Visually, there is an obvious similarity between the two. A more tenuous explanation is that the word was originally *u-nangi*. This supposedly refers to the fact that even the cormorant ("*u*"), normally most adept at catching fish, has undue difficulty ("*nangi*") in swallowing eels. A cormorant trying to swallow an eel would certainly be an amusing sight.

Eels appear in fairy tales, folklore, legends and oral traditions throughout the world, many referred to in the pages of this book, including Polynesia, Europe and Asia, as well as in local expressions or phrases (Kuroki and Tsukamoto 2012).

Fig. 6.13 An illustration in *Hokusai Manga* by the famous *ukiyo-e* artist Katsushika Hokusai in 1814, illustrating the phrase *unagi nobori* (literally "eel climbing") with its meaning of rapid, continuous increase

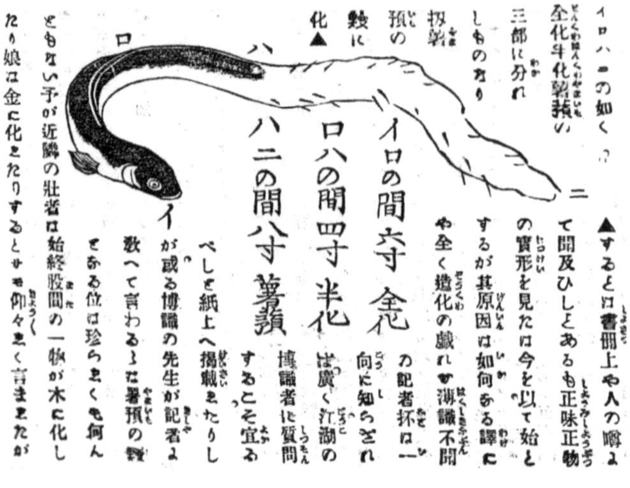

Fig. 6.14 A newspaper (*Futsū Shinbun*) published in Tokushima on 29 March 1881, reporting the catch of a strange 55 cm creature transforming from a yam to an eel, with an illustration (source, National Diet Library, Japan)

In Japanese, *unagi no nedoko* ("place where eels sleep") refers to a long passage-like room or shop with a narrow entrance, and *unagi nobori* (literally "eel climbing") describes a phenomenon that continues to arise remarkably with no apparent end. Based on the image of eels climbing vigorously up a vertical rock beside a waterfall, this idiom appears in a print produced by the famous *ukiyo-e* painter Katsushika Hokusai in the Edo era (1834). One of his *Hokusai Manga* ("Hokusai Cartoons") has the title *Unagi Nobori*, and shows three large eels escaping towards the heavens as fishers struggle desperately to prevent them from escaping (Fig. 6.13). It must have been a minor mistake by the venerable artist to show ventral fins on the eels in his print, because eels have no ventral fins!

There used to be a saying, *Yamaimo henjite unagi to naru* ("Yams turn into eels"), reflecting a general belief that eels originate from yams (*yamaimo*). In fact, not so long ago, in 1881, a newspaper even published a report about an eel transforming from a yam, illustrated by a drawing of a strange creature (a chimaera) with a half-eel, half-yam body (Fig. 6.14). This famous phrase appears in the humorous storybook *Seisuishō* ("Laughter Which Disperses Sleep"), published in the early seventeenth century. That book is said to be the origin of *rakugo* (traditional comic storytelling). In the story, a Buddhist priest cuts up an eel in the temple kitchen when one of his parishioners makes an untimely appearance. Buddhists are basically vegetarian, being prohibited from eating animals and fish, but sometimes, yielding to the temptation of appetite, they might break the precepts of their religion to eat fish secretly. Nevertheless, the priest remains unperturbed, explaining: "There is an old saying that yams turn into eels, which I used to scoff at as nonsense, but look, just when I was making some yam soup, the yam turned into an eel."

The *Man'yōshū*, Japan's oldest anthology of poetry dating back to the Nara period (710–794), famously contains two poems on the subject of eels. Both appear in Volume 16, and both were written by one of the editors, Ōtomo no Yakamochi. One of the poems concerns a colleague named Iwamaro, who was suffering from the summer heat. "Go catch some eels and eat them, then you'll feel better," the writer advises. The other poem comes with a caution: "Even if you're weak in the summer heat, at least you're still alive! Don't go falling into the river in your haste to catch an eel." The "Iwamaro" in the poem is identified as Yoshida no Muraji Oyu, an old man who was terribly thin despite having a huge appetite. The author Yakamochi is concerned about his friend's health, and advises him to eat eels, regarded as a lowly food at the time, saying that they are good for the health. He immediately follows this, but with a teasing admonition not to do so, because it is not worth risking life and limb in the dangerously fast waters of the river. The poems are titled "Two poems making fun of thin people," but in fact Yakamochi himself is said to have been quite thin. We can sense the warm wit and open-minded humour of the Man'yō era in the candid relationship between these two men.

Art and Novels

People have always expressed things familiar to them, things dear to them, in the form of pictures or carvings. For instance, images of eels can be found among those of horses, bison, mammoths and others in European cave paintings, and eels caught in the River Nile are found on Ancient Egyptian paintings and carvings. An eel can also be seen in the corner of a tapestry in the old castle of Krakow, Poland. Besides these, however, eels have been depicted in many different ways, in novels, encyclopaedias, manga cartoons, movies and animation. In this way, eels live on as part of human culture.

The Edo period in Japan saw a boom in illustrated encyclopaedic works and books on local products, some of which mention eels and are accompanied by pictures (Kuroki and Tsukamoto 2012). Among them, the illustrated encyclopaedia *Wakan Sansai Zue* written by the herbalist Terajima Ryōan, in particular, was a major work modelled on the *San Ts'ai T'u Hui* of the Ming dynasty in China. Encompassing 105 volumes in 81 books, it typifies the profuse expansion of encyclopaedic works during the Edo period. Starting with astronomy, it covers a wide range of subjects and phenomena including morality, medicine, lifestyles, performing arts, weapons, geography, and flora and fauna, with comparisons between all of these matters in China and Japan. There are of course descriptions of fish, divided into four volumes in all (freshwater and seawater, with scales and without scales, Volumes 48–51). Eels are described in Volume 50, where they are classified with catfish and loaches as freshwater fish without scales.

One of the most famous artistic expressions in the Edo era was the *ukiyo-e* print (Fig. 6.15; Kuroki and Tsukamoto 2012). *Ukiyo-e* are woodblock prints ("*e*") depicting the ways of the world ("*ukiyo*"). The motifs of *ukiyo-e* vary widely, including Kabuki (drama performance), classic stories, legends, portraits, landscapes, still life, manners and customs. Kabuki is a type of traditional Japanese drama which follows highly stylized forms and takes up stories with popular appeal. Eels were used in popular productions of Kabuki itself, or expressed in *ukiyo-e* depicting Kabuki scenes. Many *ukiyo-e* prints of popular Kabuki actors, a beautiful woman with an eel *kabayaki* dish, a famous big eel restaurant and a story from a novel remain, clearly convey the feeling of the lively megalopolis of Edo, renowned far and wide as the biggest city in the world then. Such enthusiasm for eels by people in Edo appears not only because eels were popular as food but also because people enjoyed their unique shape and humorous movements.

Eels have been the subject of several books in modern times, including Ashihei Hino's novel *Sekidō Sai* ("The Equator Festival," 1951) that was made into a movie in 1957, *Unagi no Hon* ("The Book of Eels," Isao Matsui, 1972), *Eels* (James Procek, 2010), *The Book of Eels* (T. Fort, 2002), *Consider the Eel* (Richard Shweid, 2002) and the children's

Fig. 6.15 *Ukiyo-e* prints by Utagawa Kuniyoshi: *(left) Mitate Gogyō Hi Kagaribi* (1848–1854), expressing a story from a bestselling book in the Edo era; *(centre) Ukiyo Matabei Nyōbo Otoku* (1848), dealing with the subject of a Kabuki production; *(right) Gozonji Yamakujira* (1831), depicting the popular Kabuki actor "Sawamura Tosshō" in the guise of an eel *kabayaki* chef in downtown Edo (left and centre, Ōshima-ya Collection; right, Izuei Collection)

book, *Nagahana-kun to iu Unagi no Hanashi* ("The Tale of an Eel Called Long-Nose," Konstantin Iosifov, trans. Kensuke Fukui et al., illustrated by Takaji Matsui, 1974). They also appear in the movie *Unagi* ("The Eel," director Shohei Imamura, 1997) and the popular cookery manga *Oishinbo* ("The Gourmet": Volume 3 *Sumibi no Maryoku* "The Magic of Charcoal," by Tetsu Kariya, illustrated by Akira Hanasaki, 1985), which relates various treasures of knowledge on tastes and methods of preparation. In addition, the story of a large eel taking the guise of a young woman appears in the animation *Unagisawa* "Eel Swamp" in *Manga Nihon Mukashi Banashi* ("Manga Folk Tales of Japan"). New creations continue to appear on the subject of eels themselves, or using the motif of eels. These various creations have passed the test of time before eventually becoming classics.

Eels feature in several *rakugo* stories. These include the *rakugo* classics "*Unagiya*," "*Unagi no Taiko*," "*Unagidani*," "*Shirouto Unagi*," "*Gekkyūden*" and "*Goshō Unagi*." *Unagiya* ("The Eel Broiler") is based on *Niwakatabi* ("Sudden Journey"), a story in the collection *Imayō Hanashi Kōmoku* published in 1777. An eel broiler who had recently opened a restaurant had difficulty keeping hold of his eels, and ended up chasing them through the streets. When asked by his vainly waiting customers where he was going, he delivered the famous punchline, "Don't ask me—go to the front and ask the eels!" Perhaps the humour and pathos of *rakugo* storytelling goes well with the seemingly aimless movement of eels or the rather fashionable, slightly refined atmosphere of an eel restaurant (Kuroki and Tsukamoto 2012).

Some people feel uncomfortable about eels because they resemble snakes. Nevertheless, handicrafts with eel motifs still survive in forms such as kimono sash clips, *netsuke* fasteners and clasps on tobacco pouches. Perhaps this stems from fascination with the snake-like appearance of eels, or perhaps because people find their shape attractive. Even today, there are enthusiastic fans of eels who lovingly collect miniature sets of silver eel knives, engraved eel wall hangings, Venetian glass eels, or eels made of wood, ceramics or stainless steel. Because eel skin is so tough, it has been fashioned into shoes, belts, wallets, keyholders and other useful items, although such material might come from the skin of lampreys. Eel skin has also been combined with various popular characters to be sold as souvenir keyholders and mobile telephone straps, which also sell well. Elsewhere, eels have appeared as a manga character (Unagi-inu or "Eel-Dog": Fujio Akatsuka) and a brand of biscuit ("Unagi Pie"). Finally, as a play on the phrase *unagi nobori* ("eel climbing") combined with the traditional *koi nobori* streamer, eel-shaped streamers (*unagi nobori*) and kites have been made. This wide variety of artistic expression, story or daily necessity associated with eels shows us that eels are familiar to the Japanese and loved by the people who own them.

Spirituality

Respect and Fear

There are many who think that eels are just about food. In Japan, however, some people or districts shun the idea of eating eels because they respect them as messengers of the gods, or of Buddha. There are also some areas where eating eels is considered taboo, based on a legend that eels saved their village from a big flood by filling holes in riverbanks and hence preventing water from flowing into the village (Kuroki and Tsukamoto 2012).

In many parts of Japan, people fear big old eels in ponds, swamps or deep pools, because they become "spirits of the water" after long periods of habitation. There seems to be something about eels that make us fear them as something special, drawing a clear line between eels and other fish. Possibly their unique morphology and mysterious behaviour make it easy for them be deified, giving them the image of a mysterious creature or metaphysical entity beyond human intelligence.

Votive Pictures to Shrine: *Ema*

People in Japan dedicate small wooden plaques called *ema* ("picture horses") to Shinto shrines and Buddhist temples to express prayers or gratitude for a prayer answered (Kuroki and Tsukamoto 2012). These *ema* are often painted with pictures of horses. Originally, an actual horse would be offered, but this was eventually replaced by horse-shaped objects made of wood, terracotta or other material. These were then further simplified to pictures of horses painted on plaques, producing the *ema* that survive today. Although the pictures were only of horses at first, the subjects eventually expanded to include Shinto or Buddhist deities, animals in the Chinese zodiac, or some detail relevant to the prayer being made. As such, they came to entail a strong element of folk belief. *Ema* can be seen as embodiments of ordinary people's prayers and wishes passed down through the centuries.

Eels are sometimes pictured on *ema* plaques (Fig. 6.16). Mishima Shrine in Kyoto is particularly famous for its eel *ema*. The "divine favours" sought from the shrine are fertility, safe childbirth and the prosperity of offspring. Eel *ema* can also be found at the Numanouchi Benzaiten Shrine in Iwaki, Fukushima, where fishing in the shrine's Kashikonuma Pond has been prohibited since ancient times. The pond is inhabited by eels designated as natural monuments. Antique eel *ema* still survive at Enmeiin Temple in Misato, Saitama, and Kozōji Temple in Toba, Mie. Enmeiin Temple has an *ema* consisting of two eel forks attached to a wooden board, instead of just a picture. It must have been presented as a prayer for the safety of fishers and abundant catches, as well as gratitude for past blessings—a memory of a thriving eel fishery there in days gone by.

Fig. 6.16 Eel *ema* dedicated to different shrines and temples: (*left*) Gankōzan Kozōji Temple, Toba, Mie; (*centre*) Enmei'in Temple in Misato, Saitama; (*right*) Mishima Shrine, Kyoto. Eel *ema* represent a prayer for fertility, safe childbirth and growth

Memorial Services for Eels

The Japanese consume many eels annually, but they also hold memorial services to thank the eels and to pray for their souls (Kuroki and Tsukamoto 2012). Eel mounds (Zuiganji Temple in Sendai), Gyoran Kannon Buddhas (Myōkōji Temple in Sugamo, Tokyo, Maisaka and Yoshida in Shizuoka), memorial towers and other constructions have been erected in places where these services are held, and solemn and respectful ceremonies are held in front of them. After Buddhist or Shinto ceremonies praying for their souls, eels are released into the wild. This is thought to have been modelled on the "*Hōjōe*" Buddhist ritual for releasing captured birds and fish into mountains and lakes or rivers. Scientific evaluation of these events started in recent years, to find ways of improving release techniques to enhance the eel resource.

Eels and Beliefs

Herring ovaries contain enormous numbers of eggs, which is why herring roe is known as *kazunoko* ("many children") in Japanese; herring roe is therefore a symbol of fertility and prosperity of offspring. Mice and dogs are connected with prayers for safe birth and fertility, because they give birth to many young at once. Similarly, as seen in the example of eel *ema*, the "divine favours" provided by eels are fertility, safe birth and the prosperity of offspring. However, the origin of that belief differs from that of herring or dogs. In the case of eels, the promise of favour is not based on their fecundity, because their spawning ecology was shrouded in mystery until just recently; no-one even knew how many eggs female eels could produce. Originally, it is thought that the shape of an eel's head created a link with phallus worship, and this then developed into a belief in marital harmony and fertility.

There are similar analogies, traditions and beliefs in other parts of the world, and some are mentioned elsewhere in this book. Whether in the east, the west or the tropical Pacific, eel shapes have been carved on stone rods and these then worshipped as divine spirits in exactly the same way; perhaps a form of phallicism. The Māori of New Zealand used to make greenstone pendants called *tiki*, some of which symbolized eels, and women wishing to conceive would wear them as charms in the hope that some of the eel's powers would be transferred to them. In Indonesia, it is believed that a woman not blessed with children should bathe in a lake containing eels, in Tongan folklore, there is also a legend of a young woman made pregnant by eels, and eels occasionally appear as erotic beings in the myths of Polynesia (Takayama 2009).

Taboos in Polynesia

There are many taboos involving eels in the tropical islands of Melanesia, Micronesia and Polynesia (Kuroki and Tsukamoto 2012). In Micronesia, as well as in some parts of Japan, eels are feared and not eaten (Takayama 2009). There is a legend that, long ago, a beautiful young woman spent a night with an eel, and had been eaten by the next morning. Also, there are stories of eels unexpectedly biting people's legs as they wade through shallow rivers, so eels are sometimes objects of great fear and dislike. There are also cases, on the other hand, of eels being worshipped as sacred beings, or regarded as totemic creatures from which ancestors were born. For that reason, it is considered taboo to capture or kill them, a taboo that intertwines awe, reverence and repulsion. In Tonga and Tahiti, some beliefs have eels as deities or divine messengers. It is believed that evil will transpire if these sacred eels are killed or eaten, e.g. widespread disease or the drying up of a village's precious spring water. Similar beliefs, totemism and taboos exist on various Pacific islands, and the cultures of these islands are all linked by eels.

Conclusion

The eel is clearly a special creature in Japanese history and culture. Its importance as food probably extends back as long as human history, and it is still a major form of food in Japan that is associated with national identity. Why is the consumption of eels so popular in Japan? The reason may be related to the invention of soy sauce. The grilled flavour of soy sauce has grabbed the nose and tongue of the Japanese and never let them free. Moreover, the delicacy of cooking eel fillets fostered the creation of the globally unique Japanese dish, *kabayaki*. Such common utilization of eels as food has yielded legends and idioms in society, and human feelings towards eels have been expressed in paintings and literature. Because of the closeness of eels to the Japanese people, they have become motifs in Japan for arts and crafts, fostering affection and respect. However, given their ability to inspire the counter feelings of love and hate, they have also become both taboo and images of worship or occasionally fear.

We aimed in this chapter briefly to cover the processes and systems used by Japanese to catch, rear and eat eels. However, the overexploitation of glass eels and the degradation of the freshwater environment have caused a continuing massive decline in Japanese eel numbers over the past few decades, and annual ocean–atmosphere changes offshore noted particularly of late might have resulted in further rapid decrease in the resource over the 4 years 2010–2013. To protect the Japanese eel, a common property resource in East Asia, international collaboration in resource management and research collaboration is urgently needed, and it should include international regulation of eel fisheries, maintenance of the Eel River Project of EASEC and cooperation in targeted research cruises surveying the spawning area. Further development of the techniques for artificially producing glass eels is also a critical requirement. Such endeavours will help conserve eels and in our opinion offer the greatest likelihood of sustaining the long-standing relationship between eels and humans that clearly originated as long ago as the Stone Age.

Acknowledgements We thank all the people who provided materials, artefacts, and information about the Japanese eel for this chapter, and gratefully acknowledge the assistance of Nico Korenhof (Naturalis) for scanning the originals of Keiga's watercolours and Bürger's description. We also thank Andrew Driver for his help in improving the English.

References

Aoyama J, Nishida M, Tsukamoto K (2001) Molecular phylogeny and evolution of the freshwater eel, genus *Anguilla*. Mol Phylogenet Evol 20:450–459

Chow S, Kurogi H, Mochioka N, Kaji S, Okazaki M, Tsukamoto K (2009) Discovery of mature freshwater eels in the open ocean. Fish Sci 75:257–259

Fontaine M (1936) Sur la maturation des organes de l'anguille mâle et l'émission spontanée de ses produits sexuels. Comptes Rendus de l'Académie des Sciences, vol 202. Série I, Mathématique, pp 1312–1314

Hagihara S, Aoyama J, Limbong D, Tsukamoto K (2012) Morphological and physiological changes of female tropical eels, *Anguilla celebesensis* and *Anguilla marmorata*, in relation to downstream migration. J Fish Biol 81:408–426

Ida T (2007) Unagi: Chikyū kankyō o kataru sakana (Eels: the fish that tell us about the global environment). Iwanami Shoten, Tokyo, 225 pp (in Japanese)

Inoue JG, Miya M, Miller MJ, Sado T, Hanel R, López JA, Hatooka K et al (2010) Deep ocean origin of the freshwater eel. Biol Lett 6:363–366

Kim H, Kimura S, Shinoda A, Kitagawa T, Sasai Y, Sasaki H (2007) Effect of *El Niño* on migration and larval transport of the Japanese eel, *Anguilla japonica*. ICES J Mar Sci 64:1387–1395

Kimura S, Tsukamoto K (2006) The salinity front in the North Equatorial Current: a landmark for the spawning migration of the Japanese eel (*Anguilla japonica*) related to the stock recruitment. Deep Sea Res II: Top Stud Oceanogr 53:315–325

Kimura S, Inoue T, Sugimoto T (2001) Fluctuation in the distribution of low-salinity water in the North Equatorial Current and its effect on the larval transport of the Japanese eel. Fish Oceanogr 10:51–60

Kurogi H, Okazaki M, Mochioka N, Jinbo T, Hashimoto H, Takahashi M, Tawa A et al (2011) First capture of post-spawning female of the Japanese eel *Anguilla japonica* at the southern West Mariana Ridge. Fish Sci 77:199–205

Kuroki M, Tsukamoto K (2012) Eels on the move – mysterious creatures over millions of years. Tokai University Press, Tokyo, 278 pp

Kuroki M, Aoyama J, Miller MJ, Wouthuyzen S, Arai T, Tsukamoto K (2006) Contrasting patterns of growth and migration of tropical anguillid leptocephali in the western Pacific and Indonesian Seas. Mar Ecol Prog Ser 309:233–246

Kuroki M, Miller MJ, Aoyama J, Watanabe S, Yoshinaga T, Tsukamoto K (2012) Offshore spawning for the newly discovered anguillid species *Anguilla luzonensis* (Teleostei: Anguillidae) in the western North Pacific. Pac Sci 66:497–507

Morse, E. S. 1879. Shell mounds of Omori. Memoirs of the Science Department, University of Tokio, 1: 1–36, plus 18 Plates.

Satoh H, Nakamura N, Hibiya T (1962) Studies on the sexual maturation of the eel. 1. On the sex differentiation and the maturing process of the gonads. Bull Jpn Soc Scientific Fish 28:579–584

Sugeha HY, Arai T, Miller MJ, Limbong D, Tsukamoto K (2001) Inshore migration of the tropical eels *Anguilla* spp. recruiting to the Poigar River estuary on north Sulawesi Island. Mar Ecol Prog Ser 221:233–243

Takayama J (2009) Mikuroneshia-jin ga Unagi wo Kinki suru Shūzoku no Kigen [Origin of the custom of eel taboos in micronesia]. Rokuichi Shobō, Tokyo, 220 pp (in Japanese)

Tanaka H, Kagawa H, Ohta H, Unuma T, Nomura K (2003) The first production of glass eel in captivity: fish reproductive physiology facilitates great progress in aquaculture. Fish Physiol Biochem 28:493–497

Tsukamoto K (2006) Spawning of eels near a seamount. Nature 439:929

Tsukamoto K, Nakai I, Tesch FW (1998) Do all freshwater eels migrate? Nature 396:635–636

Tsukamoto K, Chow S, Otake T, Kurogi H, Mochioka N, Miller MJ, Aoyama J et al (2011) Oceanic spawning ecology of freshwater eels in the western North Pacific. Nat Commun 2:179

Watanabe S, Aoyama J, Tsukamoto K (2009) A new species of freshwater eel *Anguilla luzonensis* (Teleostei: Anguillidae) from Luzon Island of the Philippines. Fish Sci 75:387–392

Yamamoto K, Yamauchi K (1974) Sexual maturation of Japanese eel and production of eel larvae in the aquarium. Nature 251:220–222

Yamauchi K, Nakamura M, Takahashi H, Takano K (1976) Cultivation of larvae of Japanese eel. Nature 263:412

Zenimoto K, Kitagawa T, Miyazaki S, Sasai T, Sasaki H, Kimura S (2009) The effects of seasonal and interannual variability of oceanic structure in the western Pacific North Equatorial Current on larval transport of the Japanese eel (*Anguilla japonica*). J Fish Biol 74: 1878–1890

Eels and the Korean People

7

Tae Won Lee

There are two species of freshwater eel in Korea (Kim et al. 2005): the Japanese eel (*Anguilla japonica*) and the giant mottled eel (*Anguilla marmorata*). The general distribution of the Japanese eel is compared with sea surface temperature during the glass-eel migration in Fig. 7.1 (left).

Anguilla marmorata is generally considered to be a tropical species so is rare in Korea, having been recorded in the wild only in southern Jeju Island, the northern limit of its geographic distribution. Both the species and its habitat (Fig. 7.2) have been protected under Korean law (on natural monuments) as representing one of the unique members of the country's natural fauna. However, following the relatively recent practice of importing foreign glass eels for culture, *A. marmorata* has become more common in other parts of the country and is now found in culture ponds throughout Korea, making it more difficult to maintain its protected status under law. Permission for such glass-eel imports has been sought from and granted by a national committee set up for the purpose of controlling the import of non-endemic fauna and flora. In any case, the wild area of southern Jeju is still protected under law as the natural habitat of *A. marmorata* and fishing there is strictly forbidden. Even scientific sampling can only take place there with permission, which has to be applied for from another national committee. Fortunately, *A. marmorata* has not been reported in the wild in the Korean peninsula proper, likely because even if the eels were to escape from culture ponds into rivers, they would not be able to survive the harsh winter climate.

Anguilla japonica is common throughout Korea except in the northeast, an area of the country influenced by a cold current from the north rather than by the warmer Kuroshio Current that is known to bring *A. japonica* leptocephali and thence glass eels from the south. Glass eels arrive in Korean estuaries from December to June (Fig. 7.1 right), with Jeju estuaries receiving them mainly from January to March, the Korean southern coastline from February to April, and the Korean central and western coastlines from March to May (Moon 2002). Glass eels arriving in Korean estuaries average 57.4 mm long and there are no significant differences geographically (Moon 2002). The average length is therefore slightly more than that of *A. japonica* glass eels arriving in Taiwan, southern China and Kagoshima in southern Japan, but similar to that of glass eels arriving at the main islands of Japan. The average number of growth increments in the otoliths (earbones) of the glass eels arriving in Korea is 188, with a possible but non-significant increase in number in a northward direction (Moon 2002). The number of increments is believed to provide an indication of the time in days since hatching.

Eel Biology

The sex proportions, age and growth of eels in Korean waters were summarized by Hwang (2010), who sampled them in three large estuaries, the Geum and the Mangyeng in the central west of the country, and the Nakdong on the southern coast. There are dams at the mouths of the Geum and Nakdong, but although fish ladders have been built into the dams, it is considered unlikely that eels can traverse them because they are designed for fish that swim fast, e.g. salmon.

Sex proportions of silver eels differ between estuaries (Fig. 7.3). Females dominate in the Geum and the Nakdong, but the ratio is not significantly different from unity in the Mangyeng. It is not known, however, whether the dams at the mouths of the Geum and Nakdong are the reason for the sex proportions in those estuaries being different from the undammed Mangyeng.

The ages of the silver eels caught in the different estuaries sampled by Hwang (2010) ranged between 3 and 17 years for females and between 4 and 8 years for males, with mean ages of 6 and 7 years, respectively. The mean total lengths and weights of females were also generally larger than those of males (Hwang 2010).

T.W. Lee (✉)
Department of Oceanography, Chungnam National University, Deajeon 305-764, South Korea
e-mail: twlee@cnu.ac.kr

K. Tsukamoto and M. Kuroki (eds.), *Eels and Humans*, Humanity and the Sea,
DOI 10.1007/978-4-431-54529-3_7, © Springer Japan 2014

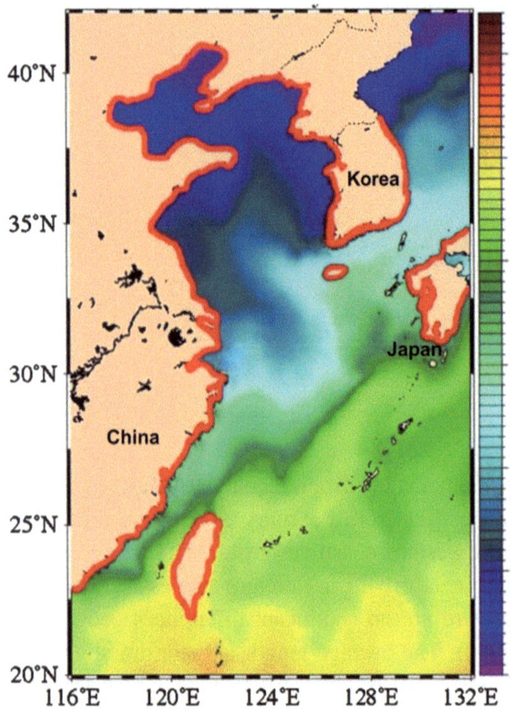

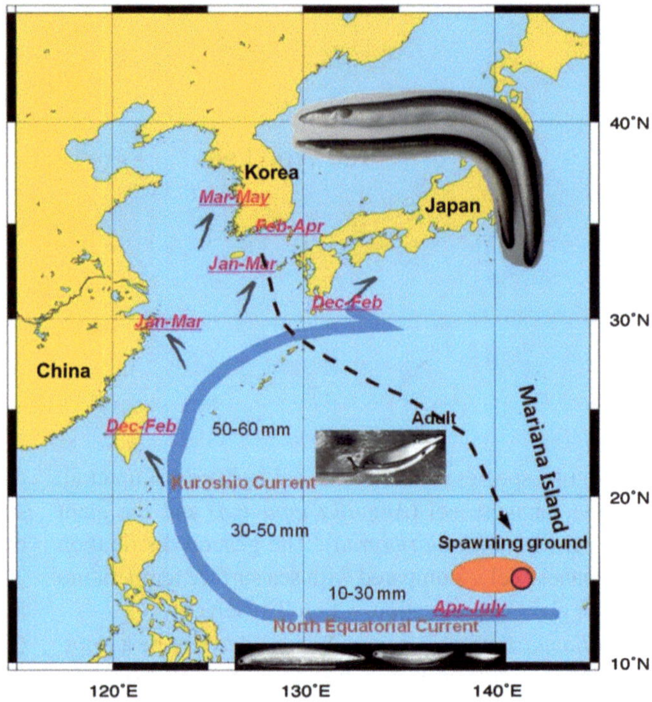

Fig. 7.1 *Left:* Geographic distribution of *Anguilla japonica* around Korea and the sea surface temperature in mid-March during the peak season for upstream migration of glass eels in Korea (SST map from http://www7320.nrlssc.navy.mil/global_ncom/plots/ecs/sst.gif). *Right:* The migration of *Anguilla japonica* with (*italicized*) the peak seasons for upstream migration of glass eels

Fig. 7.2 The area in southern Jeju protected as habitat of valuable natural fauna, in this case *A. marmorata*

Eel Production in Korea

Eel abundance declines upstream from river mouths in Korea, so eels are caught principally in estuaries. Precise statistics on eel catches are not available in Korea, but the national statistics allow at least an understanding of the trend. In the late 1960s and from the late 1970s to the late 1980s, the annual eel catch was several hundred tonnes, but since then it has dropped consistently and, since the turn of the century, it has been very small (Fig. 7.4 top). The decline in the population of *A. japonica* has, as in other countries, been

Estuaries	Female	Male	*p*
Geum River	248	6	<0.001
Mangyeong River	53	82	>0.05
Nakdong River	242	31	<0.001

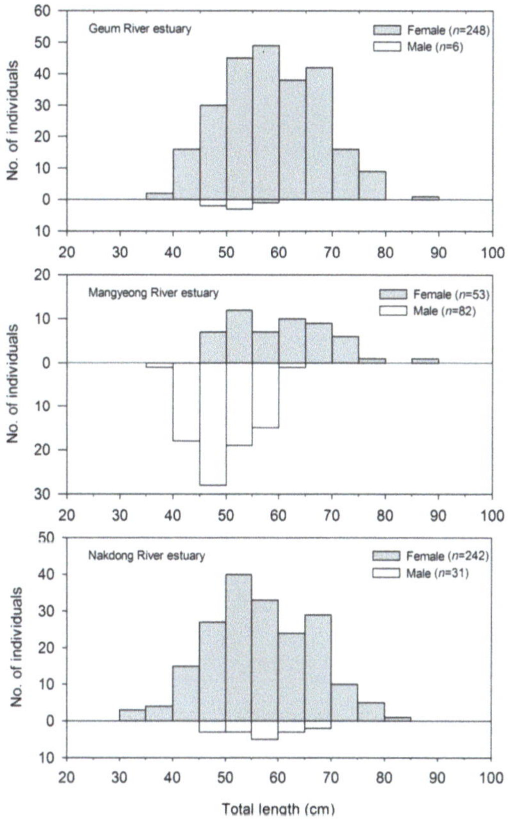

Fig. 7.3 Sex proportions and length frequency distributions by sex of eels collected from the three large river estuaries in Korea (from Hwang 2010, reproduced with permission)

variously attributed to one or all of global climate change, river pollution, dam construction and overfishing. Although internationally there is good evidence that all these issues are behind the nearly 30-year-long decline of the Japanese eel, the fact that eels are now scarce in the middle and upper reaches of rivers that have been dammed downstream, near the mouth, is widely construed in Korea to mean that the main cause of the decline has been dam construction, i.e. interrupting the eel migration upstream. By law, fish ladders have to be constructed when man-made structures across a river interfere with the movement up- or downstream of aquatic animals. Large dams have been built near the mouths of almost all the large rivers in Korea, but the fish ladders installed have almost always been designed to promote the passage of fish such as salmon that swim fast, rather than eels, which are known for their long-distance swimming ability rather than their speed. Indeed, Hwang (2010) showed that the fish ladders in many of the dams at river mouths were ineffective at allowing the passage of eels.

Almost all the current eel production in Korea is from culture, which started locally in the early 1970s. The annual production from culture is unsurprisingly well correlated to the quantity of glass eels stocked in eel ponds. It is generally known that the production in weight of a fully grown eel is approximately 800× the weight of a glass eel, so to understand the annual Korean production of eels it is necessary to investigate the catch of glass eels, or more correctly the trends in the number of glass eels stocked into culture ponds. Statistics before 1990 were obtained from www.seaworld.co.jp/~nys/chigyo.html, and the data from 1990 on from Yang Man Fish Culture Cooperative of Korea. The catch of glass eels was relatively high in the 1970s and early 1980s, and some glass eels were exported then to Japan. From the late 1980s, though, the annual domestic catch of glass eels has averaged ~7 t, but it has fluctuated between 2 and 15 t (Fig. 7.4 bottom). The importation of foreign glass eels started in the mid-1990s and exceeded ~6 t year^{-1} from 2000 on.

Eel consumption in Korea was ~7,000 t year^{-1} up to the late 1990s, but doubled after 2000, to >15,000 t year^{-1}. The surge in domestic eel consumption has been variously attributed to a decrease in the level of meat consumption as a consequence of foot-and-mouth and mad-cow disease and bird flu. Since the late 1990s too, adult eels have been imported mainly from China to meet local demand.

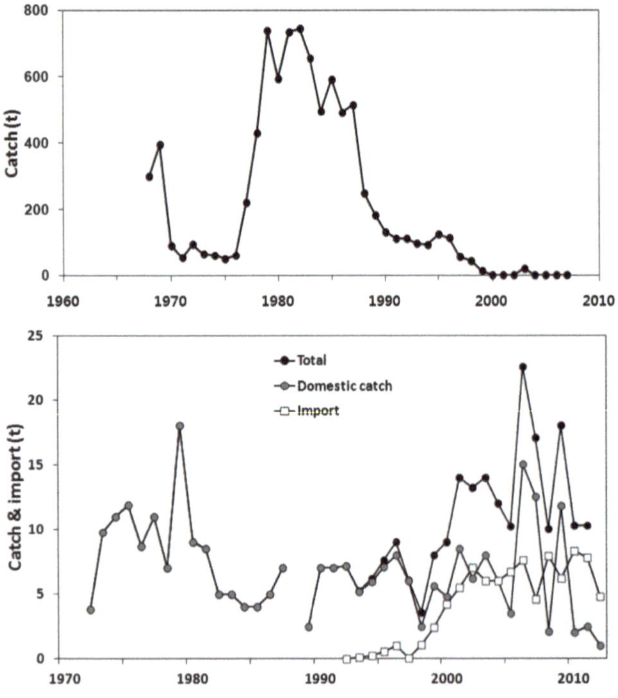

Fig. 7.4 *Top:* Annual catches of eels in the wild in Korean waters. Commercial fishing virtually stopped in 2000. *Bottom:* Annual catches, imports and total quantities of glass eels stocked in Korea

Fishing Methods

The traditional gear used to catch yellow and silver eels commercially in Korea was eel traps, longlines and bagnets. Because eels are generally scarce in the middle or upper reaches of rivers, bagnets were installed at river mouths and in estuaries to catch the migrating silver eels in autumn. Since 2000, however, directed commercial fishing for eels has virtually ceased as a result of the low density of eels in the wild. Nevertheless, eels are still taken incidentally in the bagnets installed in estuaries to catch other fish or shrimps.

Traditional small-scale methods of catching eels in Korea include spears and straw bags or gravel mounds as shelters (Fig. 7.5 top). Spearing is, however, the traditional method of catching eels worldwide, and Korea is no exception (Fig. 7.5 bottom). Here, hooked spears are dragged through the muddy bottom of estuaries using a small boat.

Glass eels have always been fished heavily in Korea, because the demand for them for culture is so high. Indeed, in Korea, with generally declining availability, glass eels have been referred to as "golden eels" simply because their value is so high and the catch relatively small. As in other

Fig. 7.5 *Top:* Catching eels at gravel mounds set as shelters for eels in the wild (photograph from http://folkency.nfm.go.kr/sesi/index.jsp). *Bottom:* Typical eel spears used in Korea, and fishers using the spears to catch eels from small boats

Fig. 7.6 Some of the methods used to catch glass eels: *from top*, scoopnets, fencenets in tidal channels, and bagnets set in tidal currents

countries, scoopnets are the preferred gear for collecting them, with fishers netting them at night under lights during flood tides (Fig. 7.6 top). Fence- and bagnets are used in western coastal waters of Korea where the tidal current is strong (Fig. 7.6 middle and bottom), with the former installed across tidal channels funnelling the glass eels into the bagnets located in the centre of wingnets. The openings of these nets face seawards, so glass eels are entrapped as they migrate from the open sea into the rivers. Bagnets are also used at river mouths, with two nets operated one each side of an anchored boat. Each net has two bars, and the net is opened by dropping the lower bar during the flood tide at night; the glass eels are then caught at high tide.

The Korean Public and Eels

Many Koreans cannot distinguish freshwater eels from marine eels such as congers and pike eels, and sometimes even hagfish are mistakenly thought to be eels. People commonly considered eels to be a fish that is dark-coloured and cylindrical. Fortunately, however, young Koreans are now taught about eels early in their school life, in primary school (Fig. 7.7), contributing to the knowledge of future generations: eels grow in freshwater, migrate to sea far from Korea, without feeding, to spawn, then the young eels return to the rivers more than 6 months old and carried there by favourable ocean currents.

Given national interest in eels, several documentaries on the species were recently aired by Korean TV networks (Fig. 7.8). KBS, the country's principal national TV network, aired two documentaries on the eel as a contribution to its topic "Special Environments" under the titles "The fish ask us where to go" and "Eels want to change." Viewers were introduced in these documentaries to the general biology and ecology of eels, and to problems relating to the migration of diadromous fish in general, especially the upstream migration of glass eels past dams. Another documentary made by MBC, a private TV network in Korea, was aired under the title "Men and Eels" as one of several topics aired on the subject "Environmental documentaries." In that programme, focus was on eel culture and the management of eel populations, but the series of programmes also highlighted the work of Japanese research expeditions on eels and eel culture in Japan. Another documentary of a similar nature was aired by FSTV, Korea's Leisure Fishing broadcasting channel. All these TV documentaries contribute to improved public understanding of eel biology and management issues.

Eels as Food

The annual catch of eels in Korea has been small relative to that of the common marine fish consumed, because eels have never been really abundant in Korean waters. As stated above, eels were rare except in the lower reaches of rivers, so were not often seen by the public inland. Eels only became more popular with the discerning Korean public once knowledge of their excellent nutritional content became well known following the start of intensive culture of the species in the 1970s. That is why the country does not have a long and historical tradition of consuming eels.

The nutritional value of eels to humankind has been emphasized during the past 40 or so years in Korea with advertisements for the fish stressing that eel meat activates the vital organs of the body and boosts energy levels. That claim was originally published in "Dong-eu-bo-gam; 東醫寶鑑," a famous medical classic in oriental countries, published in the

국어
읽기 5-2

한걸음 더
되돌아보기

문장을 구성하는 부분을 알아보았습니다. 그리고 필요한 정보를 찾으며 글을 읽어 보았습니다. 내가 잘 할 수 있는지 알아봅시다.

1. 필요한 정보를 찾으며 '뱀장어의 수수께끼'를 읽어 봅시다.

뱀장어의 수수께끼

보통 민물고기는 바다에서 살 수 없고, 바닷물고기는 민물에서 살 수 없다. 그런데 뱀장어나 연어 같은 물고기는 바닷물과 민물을 왕복하며 산다. 민물에서 자란 뒤 먼바다에서 알을 낳는 뱀장어에 대하여 알아보도록 하자.

장어는 몸이 뱀처럼 긴 물고기를 뜻하며, 우리가 흔히 볼 수 있는 장어에는 뱀장어뿐만 아니라 먹장어, 붕장어, 갯장어 등이 있다. 먹장어는 바다의 개펄 속에 주로 살며, 붕장어는 바다에서만 사는 물고기로, 뱀장어와는 다른 종류이다.

뱀장어는 민물에서 5~10년 정도 자란 뒤, 늦가을에 알을 낳으러 먼바다로 여행을 시작한다. 뱀장어가 바다로 갈 때가 되면 민물에서와는 달리 바다에서 살 수 있도록 몸이 바뀐다. 바다로 들어간 어미뱀장어가 먹이도 먹지 않고 어떻게 수천 킬로미터 떨어진 열대 태평양의 알 낳는 곳까지 6개월 이상을 여행하는지는 아직 신비에

▲ 뱀장어

Fig. 7.7 A primary school textbook telling the students about eels

early seventeenth century (Fig. 7.9). Eels have therefore been consumed as a restorative, generally prepared by boiling them with oriental medical herbs, then extracting the juice. Eels are also believed to help older folk or patients recuperating after surgery, and there are many places in Korea where one can purchase vacuum-packed "restorative eel juice" (Fig. 7.10).

Following from the burgeoning eel culture in the 1970s and the greater availability of eels thereafter, dishes of eel are now fairly common in restaurants specializing *inter alia* in such cuisine. Live eels are selected by customers and prepared directly on the customers' table, usually lightly grilled and served with a variety of sauces (Fig. 7.11), e.g. soybean (*kabayaki*), sea salt, or soy paste mixed with red pepper (*go-chu-jang*), vegetables (fresh salads or pickled cucumbers, red peppers, green perilla leaves, etc) and condiments (garlic, ginger, etc.).

Eels are eaten in Korea mainly during summer, and there are three specific days when notably nutritious foods are popular, like the unagi day at the end of July in Japan. In Korea, these days are in mid- and late July and early August. The background to these special days is that it is believed that one needs to recharge one's energy levels when hot weather has taken its toll, so eel and chicken with ginseng, both believed to be particularly nutritious dishes, are very popular then.

References

Hwang HB (2010) Sex ratio, age, growth and habitat use patterns of *Anguilla japonica* in the Geum river and Mangyeong river estuaries. Doctoral thesis, Chungnan National University, Daejeon, 95 pp

Kim IS, Choi Y, Lee CL, Lee YJ, Kim BJ, Kim JH (2005) Illustrated book of Korean fishes. Kyo-Hak Publishing Company, Seoul, 615 pp

Moon HT (2002) The early life history of eel *Anguilla japonica* determined by otolith microstructure and catch data of glass eels. Doctoral thesis, Chungnan National University, Daejeon, 111 pp

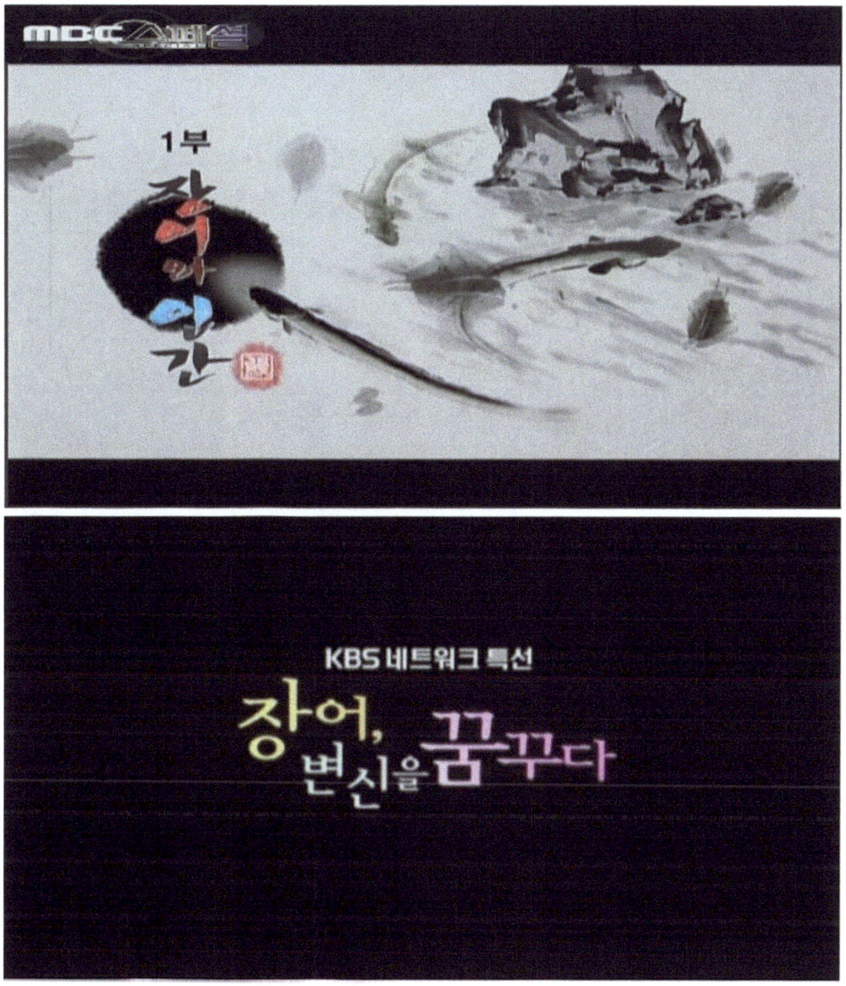

Fig. 7.8 Screenshots of the titles of two TV documentaries introducing to viewers the ecology of eels and some of the problems relating to them

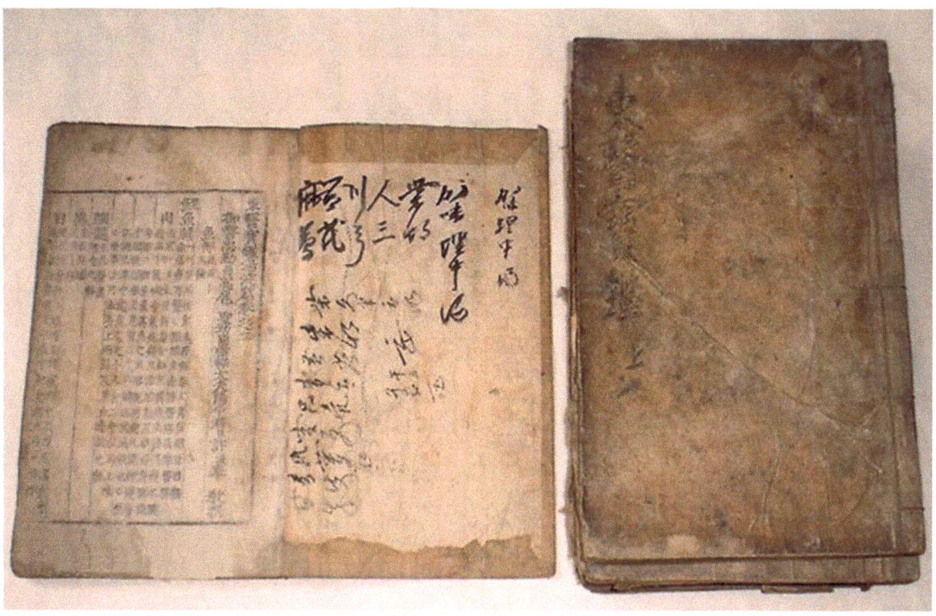

Fig. 7.9 Eels were recorded as a healthy food that reactivates vital organs and boosts energy, in "Dong-eu-bo-gam; 東醫寶鑑," a famous medical classic books in oriental countries published in the early seventeenth century

Fig. 7.10 Vacuum-packed eel juice sold in Korean shops

Fig. 7.11 Eels being prepared for customers on their tables and served with various types of condiment and vegetable

Eels in China: Species, Fisheries, Stock Management and Culture

8

Shuo-zeng Dou

Of the 165 species of eel recorded in Chinese waters (see Appendix), three (the Japanese eel *Anguilla japonica*, the conger eel *Conger japonicus*, and the pike eel *Muraenesox cinereus*) are exploited commercially or in aquaculture (especially the Japanese eel). This chapter focuses, as do the other chapters in this book, on the catadromous Japanese eel, but because of the importance of all eel species to Chinese people, information is provided on the other species too.

In the case of the Japanese eel in China, as elsewhere in its distribution, the population has been in decline for several decades, likely because of overfishing on both the incoming glass eels as well as the resident yellow eels and outwardly migrating silver eels, and as a consequence of freshwater habitat destruction caused by drought, pollution, dams and barrages and other human activities, all of which reduce the habitat available to the eels. Meanwhile, although constraints such as the dwindling glass eel supply, an unstable international market demand and upwardly fluctuating prices associated with the increasing costs of glass eels, labour and feed have impeded the development of eel aquaculture in recent years, eel aquaculture is still high-profit, so the demand for glass eels in China is increasing year on year. To date, all the glass eels used in aquaculture in China have been captured in the wild, with the price peaking at US$5 per glass eel recently, so fishers have tried to capture even more, further decimating the wild eel population. This creates a dilemma for fishery management in China and elsewhere: profit from eel aquaculture vs. sustainable wild eel stocks? These two needs have to be balanced.

As literature in the English language is not produced as widely inside China as it is outside, ready information on eels and the eel fisheries in China is relatively sparse outside the country itself. To help address this shortcoming, this chapter outlines the status of the fishery, aquaculture and stock management of commercial eel species in China, hopefully supporting increased public understanding of eels and evoking an ethos of conservation and management, especially of the endangered Japanese eel. It is hoped that the content will ensure better understanding of eel fisheries and of the human culture associated with eels in China.

Eel Species and Distribution

A total of 165 species of eel (Pisces: Anguilliformes) has been recorded in Chinese waters (Cheng and Zheng 1987; Huang 1994; Wu et al. 1999; Liu 2008; Shen and Wu 2011; Appendix). They represent 13 families: Anguillidae (3 species), Congridae (25 species), Muraenesocidae (5 species), Nettastomidae (4 species), Muraenidae (61 species), Neenchelyidae (1 species), Moringuidae (3 species), Nemichthyidae (2 species), Synaphobranchidae (14 species), Chlopsidae (2 species), Ophichthyidae (41 species), Colocongridae (2 species), Serrivomeridae (2 species). However, just four species (Japanese eel, marbled eel *A. marmorata*, conger eel and pike eel; Fig. 8.1) are considered to be fairly common throughout Chinese waters and account for almost the whole commercial catch of anguilliforms in the country. Accordingly, they are the best investigated and documented eels in Chinese literature. The other 161 species are found in the East China Sea and the South China Sea, usually shared with neighbouring Asian countries (Cheng and Zheng 1987; Nakabo 2000; Liu 2008; Shen and Wu 2011), but most are merely occasional catches made during fishery surveys or are recorded as present for taxonomic purposes; knowledge of their biology and ecology in China is therefore limited.

S.-z. Dou (✉)
Institute of Oceanology, Chinese Academy of Sciences,
7 Nan-hai Road, Qingdao 266071, People's Republic of China
e-mail: szdou@qdio.ac.cn

K. Tsukamoto and M. Kuroki (eds.), *Eels and Humans*, Humanity and the Sea,
DOI 10.1007/978-4-431-54529-3_8, © Springer Japan 2014

Fig. 8.1 The four species of anguilliform eel most commonly found in Chinese waters: (**a**) Japanese eel *Anguilla japonica*, (**b**) conger eel *Conger japonicus*, (**c**) pike eel *Muraenesox cinereus*, and (**d**) marbled eel *Anguilla marmorata*

Eel Fisheries and Aquaculture

Currently, the three primary species of anguilliform targeted by Chinese fisheries are pike eels, conger eels and Japanese eels, and between the three, a range of fishing gears, including demersal trawls, fixed traps, hooks, poles and pots are used in a range of aquatic environments (coastal waters, estuaries, lakes and streams). By total weight landed in China over the past two decades (aquaculture production excluded), pike eel catches were highest, followed by conger eel and then Japanese eel. Although not strictly the focus of this book, brief descriptions are given below of the fisheries for the first two species, before focus returns thereafter to the Japanese eel.

Pike Eel and Conger Eel Fisheries

Pike eels and conger eels are distributed widely throughout Chinese waters. Mature eels typically spawn in coastal waters between March and August, peaking from May to July. Fishing is prohibited in Chinese coastal waters between June and September, reducing the fishing pressure on spawners and hence protecting the populations and resources. Before the 1980s, the two species were taken mainly as a bycatch in trawl fisheries, but since then, targeted local fisheries with trawls, traps, hooks and poles have developed. According to official fishery statistics, the total annual landings of both species of marine eel in China have exceeded 100,000 t since 1993 and 300,000 t since 2004, with a peak of 340,000 t in 2009. Catches from the East China Sea constitute >60 % of the total pike eel landings. However, because the statistics are not comprehensive, it is not possible to determine the exact landings of each species made annually, although catches of pike eels probably account for >85 % of the combined annual catch of the two species (Zhang et al. 2007).

Despite the dramatic increase in eel landings in recent years, pike eels are in decline in Chinese coastal waters. Zhang et al. (2007) estimated the maximum sustainable yield for pike eels in the East China Sea to be ~260,000 t, less than recent annual landings, suggesting that the pike eel population there is being fully or overexploited. The current level of annual landings is therefore likely being maintained more by increased fishing effort than by a sustainable resource, so if that situation persists, future virtual collapse of the fishery may be unavoidable.

There have been a number of changes in the population structure of these two eels. For example, most of the catch of pike eels since the 1990s has been young fish (<250 mm anal length), in contrast to the situation in the 1960s and 1970s when fish 250–400 mm long dominated the catch (Zhang et al. 2007). Moreover, fish >400 mm long constituted >20 % of the catches from the 1960s to the 1980s, but <4 % of current catches. Allied to this change, the size at sexual maturity (L_{50}) of pike eels has dropped from 300 to 230 mm in females and from 210 to 160 mm in males since the 1980s, consistent with similar decreases in other overharvested fish species (e.g. croaker *Pseudosciaena polyactis* and hairtail *Trichiurus haumela*) in Chinese coastal waters. Scientists believe that this decrease in L_{50} has been caused by the combined effects of overfishing, dam and barrage construction, pollution, habitat loss and rising sea surface temperature). Long-term and significant changes in life history traits such as growth and maturation will have significant ecological consequences for eel populations and resources. Indeed, a number of fishing grounds (e.g. Haizhou Bay in the southern Yellow Sea, and Laizhou Bay in the Bohai Sea) that historically supported sizeable populations of eels have been virtually depleted since the 1990s. The present fishing grounds for these two eels are dotted around the coastal waters of Zhejiang and Fujian provinces in the East China Sea. Therefore, taken together, the observations above suggest that the populations of these commercially important anguilliform eels are in such obvious decline that reductions in the catch will have to be enforced to protect them.

Japanese Eel Fishery and Aquaculture

China has a long history of harvesting yellow and silver eels using fixed traps, hooks and pots in rivers, lakes, coastal waters and inlets. However, although the Japanese eel supports an important fishery in China, landings data are regrettably scarce, perhaps because the eels are distributed widely in freshwaters and estuaries throughout China, but at typically low biomass and with small yields at individual sites. The consequence is that the Japanese eel is not generally regarded in China as a target species. However, the lack of formal catch records does not hide the obvious decline in the eel population in recent decades. According to professional fishers, 200–500 kg of yellow eels and silver eels were being caught daily throughout autumn in large river estuaries in the 1950s and 1960s (e.g. the Liaohe, Yellow, Yangtze and Pear River estuaries). In the early 1980s, eels were still being caught, but in much lower numbers, in estuaries, rivers and lakes, but nowadays it is difficult to find wild Japanese eels in Chinese waters.

It is well documented that Japanese eel leptocephali are transported to the coastal waters of East Asia by the North Equatorial Current and the Kuroshio Current from the spawning ground west of the Mariana Islands (Tsukamoto 1992; Kimura 2003). At metamorphosis after their ocean transport, a proportion of the glass eels enters Chinese coastal waters. They first arrive in the estuaries of Fujian, Taiwan and Guangdong provinces in December, then move north to the Yangtze River estuary in late January, finally reaching the Liaohe and Yalujiang River estuaries in April (Fig. 8.2). The duration of residence varies among estuaries (e.g. 3–4 months in the Yangtze Estuary and Fujian coastal waters, 1–2 months in the Liaohe Estuary). Therefore, the seasons for glass eel fisheries vary by location. Glass eel catches are primarily used for aquaculture and to restock estuaries and river mouths, capturing the fish that might otherwise migrate upstream into rivers and recruit to the wild stocks (Fig. 8.2). Large quantities of glass eels were taken commercially before the 1980s in almost all major river estuaries in China, but the fisheries are nowadays confined to the coastal waters of Fujian province and the Yangtze River because of the depletion of the glass eel resource elsewhere. Overall, the Yangtze River estuary accounts for >80 % of the total catch of glass eels in China. Reliable statistics on the long-term trend in glass eel catches in China are elusive, but the available data do indicate that annual catches of glass eels for aquaculture during the past 5 years were >10 t, i.e. ~62.5 million individuals (at 0.16 g body weight per glass eel). Given this, it is likely that the decline in wild eel stocks is related to the increase in the fishery for glass eels in China (and elsewhere) in the late 1980s. Obviously too, increasing the harvest of glass eels for aquaculture will accelerate the decline of wild eel populations.

The earliest traceable Chinese record of eel aquaculture dates back to the Ming dynasty (1368–1644). According to the *Yang–Yu–Jing* ("Yang" means aquaculture, "Yu" means fish, "Jing" means theories, techniques and applications), an ancient encyclopaedia for fish aquaculture compiled by an agriculturist named Xingzeng Huang in the early sixteenth century, people collected "eel children" (perhaps glass eels or small yellow eels) from rivers, put them into ponds and lakes and caught them with traps when they grew. Although this may represent the start of eel aquaculture in China, intensive eel aquaculture did not begin until the mid-1970s when China started to export some of its eel production to Japan. In the 1970s, annual eel aquaculture production in China was estimated to be 23 t, but by the 1980s, this annual figure had increased to 360 t. Then, eel aquaculture production really took off, and in the 1990s, the average annual production was 92,000 t. Since then, it has been sustained at a relatively stable though slightly higher level of ~107,000 t. To support such high levels of production, large numbers of wild glass eels have to be captured every year.

Aquaculture of the Japanese eel is mainly carried out in the southern coastal provinces, including Guangdong, Fujian, Zhejiang and Shanghai (herein, "south" refers to that part of China south of the Yangtze River, which is 6,300 km long and crosses the country from western Tibet to the East China Sea). A small proportion of the production comes from inland provinces such as Jiangxi. However, throughout

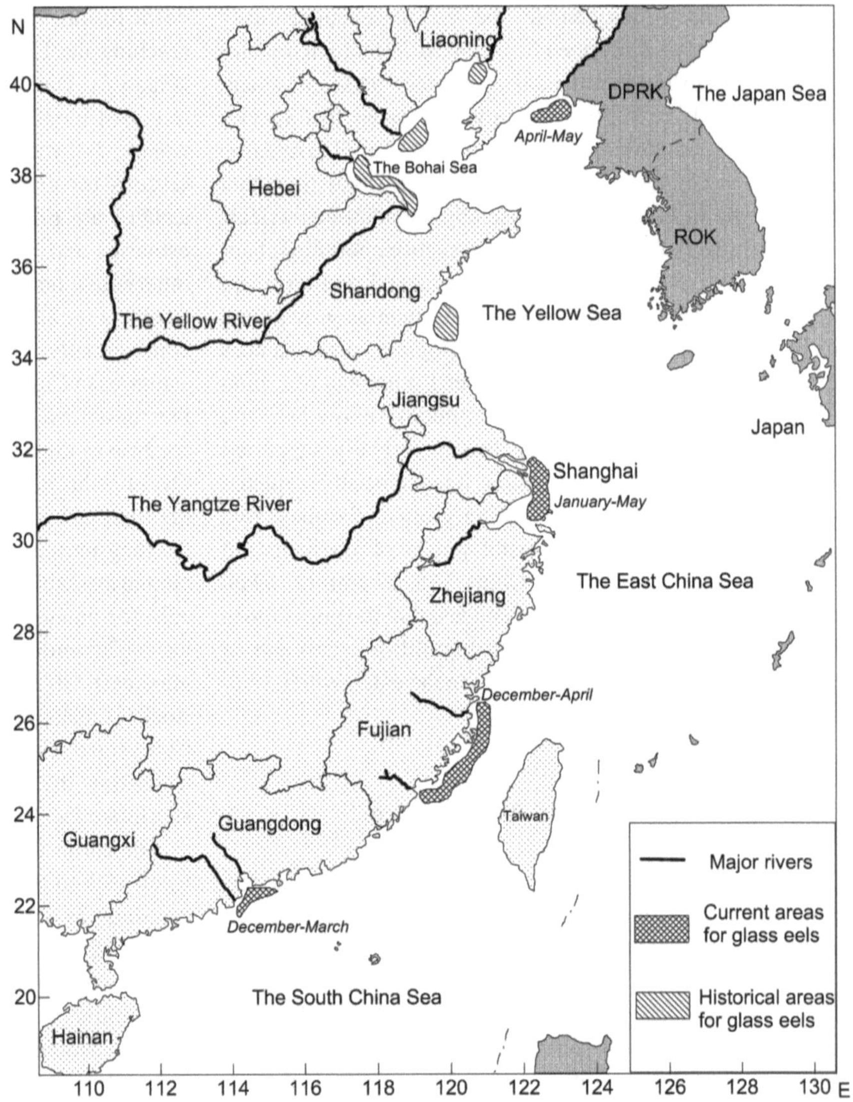

Fig. 8.2 Distribution and months of occurrence of *Anguilla japonica* glass eels along the Chinese coast

China, intensive aquaculture of eels generally takes place in indoor facilities equipped with automatic water recirculation systems, although some eels are still cultured outdoors in ponds (Fig. 8.3). More than 80 % of cultured eel product, live and processed, is exported to Japan at prices much lower than that of domestically cultured eels in Japan. In recent years, however, constraints including the decline in the glass eel supply, an unstable international market demand and price, and the increasing costs of glass eels, labour and feed have impeded the further development of eel aquaculture in China. Also noteworthy is that China started importing glass eels of the European eel *Anguilla anguilla* for aquaculture purposes, and that species has already yielded an annual production in aquaculture of >30,000 t. This mitigated the demand for glass eels of the Japanese species, but when the Convention on International Trade in Endangered Species of Wild Fauna and Flora (CITES) listed the European eel in its Appendix II as endangered in 2007 and enacted a law protecting the species in 2009, the export of European glass eels to China was banned and the practice has now ceased. Closure of that source of glass eels will undoubtedly result in further declines in the eel aquaculture industry in China. Overall, though, the decline of eel aquaculture in China will not necessarily be a bad thing for eel stocks generally; a decreasing harvest of glass eels for aquaculture might provide wild stocks with the opportunity to recover naturally.

Management of Eels in China

Because of the absence of reliable statistics and the lack of rigorously collected research data, we cannot quantify or robustly apportion blame to potential individual causes for the declines of either glass eels or yellow/silver eel popula-

Fig. 8.3 Eel fisheries and eel aquaculture in China

tions in China. However, several hypotheses exist. Increasing numbers of glass eels have been captured in estuaries for aquaculture, so the loss of those recruits from a population will have an obvious negative impact on wild eel populations. Meanwhile, freshwater habitat destruction caused by drought, pollution, the construction of dams and barrages and other human activities is relatively widespread in China, and elsewhere. Such activities reduce the habitat available to eels, so without even considering overfishing in freshwater, a combination of decreased recruitment and loss of habitat could together account for the decline in the wild eel population in China. Worse, the decline in numbers of mature eels leaving freshwater to spawn will lead to a decrease in recruitment, initiating yet another negative feedback loop impacting the species. Further, long-term climate change along with the periodic occurrence of ocean events such as *El Niño* may play an even greater role in the dramatic fluctuation and downturn in incoming glass eel numbers. The leptocephalus larvae undergo a lengthy oceanic migration, in this case within the North Equatorial Current and the Kuroshio Current, before metamorphosing into glass eels and arriving at the Chinese coast (Tsukamoto 1992; Kimura 2003). Ocean events, even if small scale during larval transport, can have a notable negative effect on eel recruitment.

The Chinese legislature, the National People's Congress (NPC), issued its first fisheries law for China (Fisheries Law of the People's Republic of China, FLPRC) in 1985 and enacted detailed rules for its implementation in 1987. The law and its rules are key to fishery management in the country (NPC 2004). Since then, regulations for fish stock management and resource exploitation have been issued and enacted in a variety of measures, covering such issues as fishing gear, closed seasons for fishing, areas closed to fishing, size at first capture and quotas for target species. Within the laws and regulations, several items are directly related to, or targeted at, eel fishery management. For example, (i) Chinese coastal waters are closed to fishing between June and September, during which period most fish species spawn and feed in coastal nursery grounds. Further, any fishing in known main spawning grounds during spawning seasons is strictly forbidden, mitigating any likelihood of overfishing the spawners. (ii) Electrofishing and the use of explosives are both banned in rivers, lakes, streams and estuaries to prevent freshwater species and migratory species such as the Japanese eel from being overharvested or their habitat destroyed. (iii) The inclusion in hydraulic constructions such as weirs, barrages and dams of a fish passage for migratory species is compulsory. (iv) Fishing the spawners of overexploited and depleted fishery stocks (e.g. Japanese eel, tapertail anchovy *Coilia nasus*, Chinese sturgeon *Acipenser sinensis*, and shad *Hilsa reevesi*) is prohibited in both rivers and estuaries in an effort to increase the levels of recruitment. (v) A minimum size at first capture is stipulated for many target species, and appropriate minimum mesh sizes for the gear used in each fishery are imposed to ensure that young fish are not being harvested. Recently, the appropriate mesh size by gear and the size at first capture has been determined for >30 target fish species. For pike eels for instance, the length and the age at first capture are 340 mm and 5–6 years, respectively (Zhang et al. 2007).

As mentioned above, *A. japonica* glass eels are massively overharvested in China, attempting to meet the burgeoning demands of aquaculture. Moreover, a reasonable proportion of the small yellow eel population has to be allowed to migrate upstream in rivers to maintain and grow the wild populations, so a range of measures has been implemented to balance competing needs. For example, the capture of glass eels is generally limited to a period of ~2 months after their arrival in a specific estuary. As glass eels may occupy an estuary for 3–4 months, a reasonable portion should escape capture and hence have the chance to migrate upstream in the rivers. In addition, attempting to capture young eels in rivers is prohibited, providing opportunity for them to recruit into the wild populations every year. Other measures of protecting the species and maintaining effort at reasonable levels include strict quota controls and limits on the number of active fishing licences. Restocking glass eels and yellow eels into rivers to enhance wild eel populations has also been attempted recently, typically overseen by local government fishery managers, but the programme is still at an early, experimental stage and small scale, so evidence for any positive effect of the activity is still elusive.

Eel aquaculture is a high-profit industry and the demand for Japanese glass eels increases annually in China. Although artificially cultured hatchery glass eels can now be obtained using techniques recently developed in Japan (see elsewhere in this book), there is still a long way to go before sufficient numbers of hatchery eel fry will be produced to support commercial aquaculture at current levels. As stated above, all the glass eels for aquaculture in China are still captured in the wild, and the value to the fishers catching them is now so high that they are encouraged to catch as many as they can, driving stocks down even further. The answer is obvious to ecologists and conservationists alike: if the intent is to conserve the eel population in China and elsewhere, eel aquaculture using wild-caught glass eels has to be regulated at a low level, much lower than currently exists. If that does not happen soon, the future for the species, let alone the industry based on it in China, is dire.

Eels and Human Culture

In ancient China, people apparently knew of and utilized eels primarily for medicinal purposes rather than as food. They believed that eels produced with various herbs (Chinese medicines) could have positive medical effects on a range of ailments, including haemorrhoids, beri-beri, kidney problems and pulmonary tuberculosis, and that eels also acted as an aphrodisiac. According to the Chinese medical classic *Compendium of Materia Medica*, compiled by a folk doctor named Shizhen Li in 1552, eels boiled with wine, vinegar, herbs (herbaceous peony root, rehmannia, angelica) and turtle shell could help in the treatment of tuberculosis and vitiligo (a condition that causes depigmentation of sections of skin that occurs when melanocytes, the cells responsible for skin pigmentation, die or cannot function). Many of the medical recipes prescribed in that classic have been proven in contemporary medical research to be both reasonable and scientific. Unfortunately, however, there is no scientific evidence to confirm that it is the eels rather than the herbs that have the medical benefit described in the book, although the text does suggest that Chinese people have utilized eels in their daily lives for a long time.

Folk legends in China also feature ancient beliefs about eels. One legend goes like this: long long ago, a village named Huanglian–lin ("coptis woods") was plagued with a strange disease (actually tuberculosis). People died of the disease one by one, but no-one could find an effective medicine to stop the spread of the pestilence. One day, a prophet came to the village and told the people that they were cursed by a river devil because he had asked for a young maiden from the village as his bride, but the villagers had not agreed to provide him one. It was clear, therefore, that someone had to be sacrificed to the devil so that the plague could be contained, a harsh choice for anyone to make. However, Chunni (meaning "spring maid"), a seriously ill 18-year-old maiden, volunteered to go to the devil to save all the other villagers. Sadly therefore, the villagers loaded Chunni onto a small boat and allowed it to drift out into the river. After a long journey of many days, the boat reached the estuary and a kind-hearted young man named Wang Xiaofa found the dying maiden there, took her to his home in the appropriately named "eel village" (all people in it making a living by fishing for eels) and fed her eels daily. Miraculously, the maiden recovered, went back with her young man to her village and saved all the other sick villagers' lives with the eels they caught. After that, the young couple were married and lived happily ever after, having a large family that generated a strong line of successors. Although no-one really knows whether this story is true or pure fiction, the point is that the legend does reflect how important eels were to ancient Chinese people and their culture.

In the Chinese language, eels also feature as proverbs and metaphors. For example, "the water snake attacks the eel—to see who is longer or stronger" means that, in a fight or competition, power and strength decide the winner. It also means that all opponents will be seriously hurt in a fight or competition because they are strong enough to attack each other. Another example, "the loach meets the eel in a muddy paddy—the right rivals" means that slippery fellows challenge each other and it is hard to judge who is the winner because both are sly and not easily defeated. Of course, the metaphor "marriage is like an eel trap" is not difficult for all married people to understand. Eels therefore remind ancient Chinese people of a snake, an animal that is usually a metaphor for slyness, cruelty and infidelity, possibly because they share a long, sleek, viscous body as well as a preference for living in dark burrows. Succinctly, eels seem to be unfairly treated as a metaphor for bad things or people in Chinese folklore.

Despite their unfortunate reputation since ancient times, eels today enjoy considerable popularity as a delicacy in China. Besides popular Japanese eel foods such as eel *sushi* (rice wrapped with toasted conger eels and Japanese eels) and *kabayaki* (barbecued eel served with special sauces), there are many other forms of eel cuisine in China, including fried, stewed, steamed and dried eels, and eel soup (Fig. 8.4). Different eel foods typically require different accompanying herbs (e.g. pepper, onion, aniseed, garlic, ginger, hot pepper) and sauces (e.g. vinegar and wine) for appropriate flavouring and seasoning. They might be cooked in different ways, but they share a common theme: delicacy. Generally, eels are consumed more in the southern than in the northern part of China. People are served eels in speciality restaurants, but they also enjoy them at family dinner tables because the recipes are relatively simple to follow and widely available.

In recent years, cultural eel festivals have been held in autumn in many parts of China, particularly in the southern Chinese provinces of Zhejiang, Fujian and Guangdong, where large quantities of Japanese eels are still produced (wild and aquaculture) and consumed. The festivals are usually organized by local eel fishery associations and are widely attended by people from surrounding districts. In addition, such festivals are sometimes supported and attended by local government fishery managers. The theme generally celebrates the preservation of traditional Chinese eel food culture, and the harvest, and promotes and advertises the principles of sustainable exploitation, aquaculture, enhancement and protection of eel resources. Such activities increase public understanding of and knowledge about eels and promote a conservation ethic for this clearly endangered species.

Fig. 8.4 Typical eel dishes in China

Appendix

Eel species of China (sources: Cheng and Zheng 1987; Huang 1994; Liu 2008; Wu et al. 1999; Shen and Wu 2011). *BS* Bohai Sea, *YS* Yellow Sea, *ECS* East China Sea, *SCS* South China Sea

			Distribution			
Family	Genus	Species	BS	YS	ECS	SCS
Anguillidae	*Anguilla*	*A. japonica* Temminck and Schlegel, 1847	+	+	+	+
		A. marmorta Quoy and Gaimard, 1824		+	+	+
		A. bicolor pacifica Schmidt, 1928			+	+
Congridae	*Conger*	*C. cinereus* Rüppell, 1830			+	+
		C. japonicus Bleeker, 1879		+	+	
		C. myriaster (Brevoort, 1856)	+	+	+	
	Coloconger	*C. scholesi* Chan, 1967				+
	Ariosoma	*A. anago* (Temminck and Schlegel, 1846)			+	+
		A. anagoides (Bleeker, 1854)			+	+
		A. major (Asano, 1958)			+	+
		A. fasciatus (Richardson, 1846)			+	+
		A. nancyae Shen, 1998			+	+
	Bathymyrus	*B. simus* Smith, 1965			+	+
	Bathyuroconger	*B. vicinus* (Vaillant, 1888)			+	+
	Bathycongrus	*B. gattulatus* (Günther, 1887)			+	+
		B. retrotinctus (Jordan and Snyder, 1901)			+	+
		B. wallacei (Castle, 1968)			+	+
	Gnathophis	*G. nystromi nystromi* (Jordan and Snyder, 1901)				+
		G. xenica (Matsubara and Ochiai, 1951)			+	+
	Gorgasia	*G. japonica* Abe, Miki and Asai, 1977			+	+
		G. taiwanensis Shao, 1990			+	+
	Heteroconger	*H. hassi* (Klausewitz and Eibl-Eibesfeldt, 1959)			+	+
	Rhynchoconger	*R. brevirostris* Chen and Weng, 1967			+	+
		R. ectenurus (Jordan and Richardson, 1909)			+	+
	Parabathymyrus	*P. macrophthalmus* Kamohara, 1938			+	+
	Uroconger	*U. lepturus* (Richardson, 1845)			+	+
	Congriscus	*C. megastomus* (Günther, 1887)			+	+
	Japonoconger	*J. sivicolus* (Matsubara and Ochiai, 1951)			+	+
Muraenesocidae	*Muraenesox*	*M. bagio* (Hamilton, 1822)			+	+
		M. cinereus (Forsskål, 1775)	+	+	+	+
	Congresox	*C. talabon* (Cuvier, 1829)			+	+
		C. talabonoides (Bleeker, 1853)			+	+
	Oxyconger	*O. loptognathus* Bleeker,1867			+	+
Nettastomidae	*Nettastoma*	*N. parviceps* Günther, 1877			+	+
		N. taiwanensis (Chen and Weng, 1967)			+	
	Saurenchelys	*S. fierasfer* (Jordan and Snyder, 1901)			+	+
		S. taiwanensis Karmovskaya, 2004			+	+
Muraenidae	*Anarchias*	*A. allardicei* Jordan and Starks, 1906			+	+
		A. cantonensis (Schultz, 1943)			+	+
		A. fuscus Smith, 1962			+	+
	Cirrimaxilla	*C. formosa* Chen and Shao, 1995			+	+
	Echidna	*E. delicatula* (Kaup, 1856)			+	+
		E. nebulosa (Ahl, 1789)			+	+
		E. polyzona (Richardson, 1845)			+	+
		E. xanthospilos (Bleeker, 1859)			+	+
	Rhinomuraena	*R. quaesita* Garman, 1888			+	+
	Strophidon	*S. brummeri* (Bleeker, 1858)			+	+
		S. sathete (Hamilton, 1922)			+	+
	Gymnomuraena	*G. zebra* (Shaw and Nodder, 1797)			+	+
	Uropterygius	*U. concolor* Rüppell, 1838			+	+
		U. macrocephalus (Bleeker, 1865)				+
		U. marmorata (Lacepède, 1803)			+	+
		U. micropterus (Bleeker, 1852)			+	+

(continued)

Appendix (continued)

Family	Genus	Species	Distribution BS	YS	ECS	SCS
		U. nagoensis Hatooka, 1984			+	+
		U. tigrinus (Lesson, 1828)			+	+
	Enchelycore	*E. bayeri* (Schultz, 1953)			+	+
		E. bikiniensis (Schultz, 1953)			+	+
		E. lichenosa (Jordan and Snyder, 1901)			+	+
		E. pardalis (Temminck and Schlegel, 1846)			+	+
		E. schismatorhynchus (Bleeker, 1853)			+	+
	Gymnothorax	*G. albimarginatus* (Temminck and Schlegel, 1846)			+	+
		G. berndti Snyder, 1904			+	+
		G. buroensis (Bleeker, 1857)			+	+
		G. chilospilus Bleeker, 1865			+	+
		G. chlamydatus Snyder, 1908			+	+
		G. eurostus (Abbott, 1861)			+	+
		G. favagineus (Bloch and Schneider, 1801)			+	+
		G. fimbriata (Bennctt, 1832)			+	+
		G. flavimarginatus (Rüppell, 1830)			+	+
		G. formosus Bleeker, 1865			+	+
		G. hepaticus (Rüppell, 1830)			+	+
		G. herrei Beebe and Tee-Van, 1933			+	+
		G. javanicus (Bleeker, 1859)			+	+
		G. kidako (Temminck and Schlegel, 1846)			+	+
		G. leucostigma Jordan and Richardon, 1909			+	+
		G. margaritophorus Bleeker, 1865			+	+
		G. melanospilus (Bleeker, 1855)			+	+
		G. meleagris (Shaw and Nodder, 1795)			+	+
		G. monostigmus (Regan, 1909)			+	+
		G. neglectus Tanaka, 1911			+	+
		G. niphostigmus Chen, Shao and Chen, 1996			+	+
		G. nudivomer (Günther, 1867)			+	+
		G. pindae Smith, 1962)			+	+
		G. pictus (Ahl, 1789)			+	+
		G. polyuranodon (Bleeker, 1853)			+	+
		G. prionodon Ogilby, 1895			+	+
		G. prolatus Sasaki and Amaoka, 1991			+	+
		G. pseudothyrsoideus (Bleeker, 1852)			+	+
		G. punctatofasciatus Bleeker, 1863			+	+
		G. reevesi (Richardson, 1845)				+
		G. reticularis Bloch, 1795			+	+
		G. richardsoni (Bleeker, 1852)				+
		G. rüppelliae (McClelland, 1844)			+	+
		G. thysoideus (Richardson, 1845)			+	+
		G. undulatus (Lacepède, 1803)			+	+
		G. ypsilon Hatooka and Randall, 1992			+	+
		G. zonipectis Seale, 1906			+	+
	Thyrsoidea	*T. macrurus* (Bleeker, 1854)			+	+
Neenchelyidae	*Neenchelys*	*N. parvipectoralis* Chu, Wu and Jin, 1981			+	+
Moringuidae	*Moringua*	*M. abbreviata* (Bleeker, 1865)			+	+
		M. macrocephalus (Bleeker, 1865)			+	+
		M. maccrochir Bleeker, 1855			+	+
Nemichthyidae	*Avocettina*	*A. infans* (Günther, 1878)			+	+
	Nemichthys	*N. scolopaceus* Richardson, 1848			+	+
Synaphobranchidae	*Dysomma*	*D. anguillaris* Barnard, 1923			+	+
		D. dolichosomatus Karrer, 1982			+	+
		D. goslinei Robins and Robins, 1976			+	+
		D. melanurum Chen and Mok, 1995			+	+

(continued)

Appendix (continued)

Family	Genus	Species	Distribution BS	YS	ECS	SCS
		D. opisthoproctus Chen and Mok, 1995				
		D. polycatodon Karrer, 1982			+	+
	Dysommina	*D. rugosa* Ginsburg, 1951			+	+
	Meadia	*M. roseni* Mok, Lee and Chan, 1991			+	+
	Simenchelys	*S. parasiticus* Gill, 1879			+	+
	Synaphobranchus	*S. affinis* Günther, 1877			+	+
		S. brevidorsalis Günther, 1887			+	+
		S. kaupii Johnson, 1862			+	+
		S. pinnatus (Gronow, 1854)			+	+
	Histiobranchus	*H. bathybius* (Günther, 1877)			+	
Chlopsidae	*Kaupichthys*	*K. diodontus* (Schultz, 1943)			+	+
	Chilorhinus	*C. platyrhynchus* (Norman, 1922)			+	+
Ophichthyidae	*Caecula*	*C. longipinnis* (Kner and Steindachner, 1867)			+	+
	Bascanichthys	*B. kirkii* (Günther, 1870)			+	+
	Callechelys	*C. catostomus* (Schneider and Forster, 1801)			+	+
		C. maculates Chu, Wu and Jin, 1981			+	+
		C. marmorata (Bleeker, 1853)			+	+
	Brachysomophis	*B. cirrhochilus* (Bleeker, 1859)			+	+
		B. crocodilinus (Bennett, 1833)			+	+
	Cirrhimuraena	*C. chinensis* Kaup, 1856			+	+
		C. yuanding Tang and Zhang, 2003			+	+
	Echelus	*E. uropterus* (Temminck and Schlegel, 1846)			+	+
	Leiuranus	*L. semicinctus* (Lay and Bennett, 1839)			+	+
	Myrichthys	*M. colubrinus* (Boddaert, 1781)			+	+
		M. maculosus (Cuvier, 1817)			+	+
	Myrophis	*M. cheni* Chen and Weng, 1967			+	+
	Ophichthus	*O. aphotistos* McCosker and Chen, 2000			+	+
		O. apicalis (Anonymous, 1830)			+	+
		O. asakusae Jordan and Snyder, 1901			+	+
		O. bonaparti (Kaup, 1856)			+	+
		O. brevicaudatus Chu, Wu and Jin, 1981			+	+
		O. celebicus (Bleeker, 1856)				+
		O. cephalozona (Bleeker, 1864)			+	+
		O. erabo (Jordan and Snyder, 1901)			+	+
		O. evermanni Jordan and Richardson, 1909			+	+
		O. fasciatus (Chu, Wu and Jin, 1981)			+	+
		O. macrochir (Bleeker, 1853)			+	
		O. polyophthalmus Bleeker, 1856			+	+
		O. stenopterus (Cope, 1871)			+	
		O. tchangi Tang and Zhang, 2002			+	
		O. tsuchidae Jordan and Snyder, 1901			+	
		O. urolophus (Temminck and Schlegel, 1846)			+	+
	Pisodonophis	*P. boro* (Hamilton-Buchanan, 1822)			+	+
		P. cancrivorus (Richardson, 1844)			+	+
		P. rubicandus Chen, 1829			+	+
		P. zophistius Jordan and Snyder, 1901			+	+
	Mystriophis	*M. porphyreus* (Temminck and Schlegel, 1845)			+	
	Ophisurus	*O. macrorhynchus* Bleeker, 1853			+	
	Xyrias	*X. revulsus* Jordan and Snyder, 1901			+	+
	Muraenichthys	*M. gymnopterus* (Bleeker, 1853)			+	+
		M. gymnotus Bleeker,1857			+	+
		M. macropterus Bleekr, 1857			+	+
		M. malabonensis Herre, 1923				+
Colocongridae	*Coloconger*	*C. japonicus* Machida, 1984			+	+
		C. scholesi Chan, 1967			+	+
Serrivomeridae	*Serrivomer*	*S. beani* Gill and Ryder, 1883				+
		S. sector Garman, 1899			+	

References

Cheng QT, Zheng BS (eds) (1987) Fishes of China with pictorial keys to the species. Science Publishing Press, Beijing, 1458 pp (in Chinese)

Huang ZG (ed) (1994) Marine species and their distribution in China. Ocean Publishing Press, Beijing, 754 pp (in Chinese)

Kimura S (2003) Larval transportation of the Japanese eel. In: Aida K, Tsukamoto K, Yamauchi K (eds) Eel biology. Springer, Tokyo, pp 169–179, 497 pp

Liu JY (ed) (2008) Checklist of marine biota of China seas. Science Publishing Press, Beijing, 1267 pp (in Chinese)

Nakabo T (ed) (2000) Fishes of Japan with pictorial keys to the species. Tokai University Press, Tokyo, 1748 pp (in Japanese)

NPC (2004) Fisheries law of the People's Republic of China (FLPRC) and the detailed rules for implementation. China Democracy and Law Press, Beijing, 21 pp (in Chinese)

Shen SJ, Wu GY (eds) (2011) Fishes of Taiwan. Yueya Commercial Design Company, Taipei, 896 pp

Tsukamoto K (1992) Discovery of the spawning area for Japanese eel. Nature 356:789–791

Wu HL, Shao GZ, Lai CF (eds) (1999) Latin–Chinese dictionary of fishes names. Fisheries Publishing Press, Taipei, 1028 pp (in Chinese)

Zhang QH, Cheng JH, Xu HX, Shen XQ, Yu GP, Zheng YJ (eds) (2007) Sustainability of biological resources in the East China sea. Fudan University Press, Shanghai, 731 pp (in Chinese)

Freshwater Eels and Humans in Taiwan

9

Wann-Nian Tzeng

Freshwater eels (*Anguilla* spp.) are an important food resource and support large-scale aquaculture in some oriental countries. Taiwan's climate is ideal for eel stocks living in the wild and for aquaculture. Indeed, the country's aquaculture industry, which was initiated in the 1960s and peaked in the early 1990s, contributed greatly to Taiwan's economic development at the time, though much of that production has since been transferred to mainland China. Despite their economic importance to humans, however, many people are not that familiar with eels, so this chapter supplements material presented elsewhere in this book by documenting aspects of eel distribution and biology, mythology, cuisine, etc, in Taiwan.

Eel Species in Taiwan

Of 19 species or subspecies of freshwater eel known worldwide (Watanabe et al. 2009), five have been reported in Taiwan, the Japanese eel (*Anguilla japonica*), the giant mottled eel (*A. marmorata*), *A. bicolor pacifica*, *A. celebesensis* and *A. luzonensis*. All except the last of these have been reported for some time as living naturally in Taiwan (Tzeng 1982, 1983; Tzeng and Tabeta 1983; Han et al. 2001), but relatively recently, two new species were identified by molecular analysis, *A. luzonensis* in a river of northern Luzon, the Philippines (Watanabe et al. 2009) and *A. huangi* at an aquaculture farm in Taiwan to which glass eels had been brought from northern Luzon for aquaculture (Teng et al. 2009). Those two new species are, however, synonymous (Leander et al. 2012), so according to Article 23 of the International Code for Zoological Nomenclature, the first nomenclatural act or the first published name receives precedence. In the case of the two new species, the name *luzonensis* was published in March 2009 and *huangi* in November 2009, so the species name *luzonensis* has priority. Further, the identification of *A. celebesensis*, described by Tzeng (1982), was placed in doubt because of the similarity of its external morphology with that of *A. luzonensis*, but with the recent simultaneous identification of both species as separate by DNA analysis (Chang et al. 2012), the two species have been maintained. Accordingly, it is confirmed that five species of *Anguilla* are found in Taiwan.

The colour pattern is uniform in *A. japonica* and *A. bicolor pacifica* and marbled in *A. marmorata, A. celebesensis* and *A. luzonensis. A. marmorata* can grow to >170 cm total length (Fig. 9.1), but the more common *A. japonica* does not reach that length in the wild. *A. bicolor pacifica* is a short-finned eel, the other four are longfinned, *A. japonica* is classified as temperate, and the other four as tropical (Tesch 2003). The geographic distribution of the five species differs too, with *A. japonica* of the two more common species dominating the lower reaches of rivers and *A. marmorata* preferring deep pools in upper reaches (Shiao et al. 2003). The other three species are rare (Tzeng et al. 1995). Clearly, though, the eels in Taiwan, their morphology, behaviour and habitat use, and the environment they inhabit, are diverse.

The Japanese eel is the most abundant of the five species known in Taiwan. Its glass eels recruit naturally to Taiwanese estuaries during winter, from November to February, and have traditionally been fished during their upstream migration to be ongrown in aquaculture farms (Tzeng 1985). Daily growth increments in the otolith (the ear stone) have revealed that Japanese eel leptocephalus larvae take some 120–140 days to arrive at the estuaries of Taiwan, mainland China, Korea and Japan (Cheng and Tzeng 1996), after the adult eels have spawned during the new moon of May/June west of the Mariana Islands (Tsukamoto 1992), the leptocephali being transported by the North Equatorial Current (NEC) and the Kuroshio Current (see the Japanese chapter in this book for more detail). However, despite anguillid eels being

W.-N. Tzeng (✉)
Institute of Fisheries Science, National Taiwan University, Taipei 10617, Taiwan

Department of Environmental Biology and Fisheries Science, National Taiwan Ocean University, Keelung 20224, Taiwan
e-mail: wnt@ntu.edu.tw

K. Tsukamoto and M. Kuroki (eds.), *Eels and Humans*, Humanity and the Sea,
DOI 10.1007/978-4-431-54529-3_9,

Fig. 9.1 Colour patterns of *A. japonica* (*top left*) and *A. marmorata* (*bottom left*). The latter can grow to >170 cm (*right*, from Williamson and Boëtius 1993)

classified generically as catadromous, proven too by analysis of strontium/calcium (Sr:Ca) ratios in the otolith (Tzeng et al. 2000, 2003; Tsukamoto and Arai 2001; Thibault et al. 2007), part of the Japanese eel population completes its whole life cycle in seawater or at least in very brackish water.

The recent decline in the population of wild Japanese eels has resulted in there being an insufficient supply of glass eels for aquaculture in Taiwan and elsewhere. The reason for the decline is not absolutely clear, but as speculated elsewhere in this book and for other species of eel too, it may be related *inter alia* to overfishing, habitat degradation and/or global climate change (see below). Whatever the cause of the decline, though, and in an attempt to stimulate recovery of Japanese eels in the wild and concomitantly to increase glass eel production for aquaculture, the Taiwanese government ordered the release of hormone-induced mature eels (silver eels) into the open ocean from 1976 to 2002. Since the millennium, however, that programme has shifted its focus to releasing young eels into rivers.

In addition to the five species of eel found naturally in Taiwan, some exotic species of anguillid eel have also been found in the wild. Succinctly, faced with a reducing inflow of Japanese glass eels and a heavy demand for glass eels generally for aquaculture in the country, glass eels of non-endemic species such as the American eel *A. rostrata* were introduced from North America; some escaped from the aquaculture ponds into the wild and have since been caught occasionally during their spawning migration as adults (Han et al. 2002; Tzeng et al. 2009). Additionally, the Australian speckled longfin eel *A. reinhardtii* has been caught in Sun-Moon Lake in central Taiwan, having originally been imported from Australia for cuisine purposes because of its similarity to the *A. marmorata* eaten preferentially by Taiwanese (Chang et al. 2008).

Glass Eels

The species composition of glass eels differs between the eastern and the western coasts of Taiwan (Fig. 9.2). In the Hsiukuluan River estuary of eastern Taiwan (*A. marmorata* glass eels dominate to the tune of 97.1 % of the total catch and the other three species are scarce: *A. bicolor pacifica* (1.17 %), *A. japonica* (0.5 %), *A. luzonensis* (1.2 %). In contrast, *A. japonica* glass eels dominate along the northwestern coast of Taiwan, making up 81.0 % of the total glass eel catch, with *A. marmorata* (17.8 %), *A. bicolor pacifica* (1.0 %) and *A. luzonensis* (0.2 %) less or much less abundant. In the Donkang River estuary of southwestern Taiwan, the proportional representation by species of glass eel is between these two extremes: *A. marmorata* is most abundant (64.9 %), followed by *A. japonica* (29.7 %), *A. bicolor pacifica* (4.7 %) and *A. luzonensis* (0.7 %). The differences in geographic distribution and abundance among species around Taiwan are attributable to the spawning site of origin as well as the habitat (tropical or temperate) and the ocean currents around the island (warm Kuroshio Current or cold China Coastal Current) that transport the eel larvae in towards the coasts (Cheng and Tzeng 1996; Han et al. 2012; Leander et al. 2012).

Long-term catch data (1972–2011) have indicated a significant decadal change in peak catches of Japanese glass eels coinciding with solar activity reflected in an 11.2 year periodic change in sunspot number (Fig. 9.3a; Tzeng et al. 2012a). The catch of glass eels seems to increase with a concomitant increase in the number of sunspots, and although the cause–effect relationship between glass eel numbers and sunspots is not a direct one, the climate change index WPO (Western Pacific Oscillation) that influences

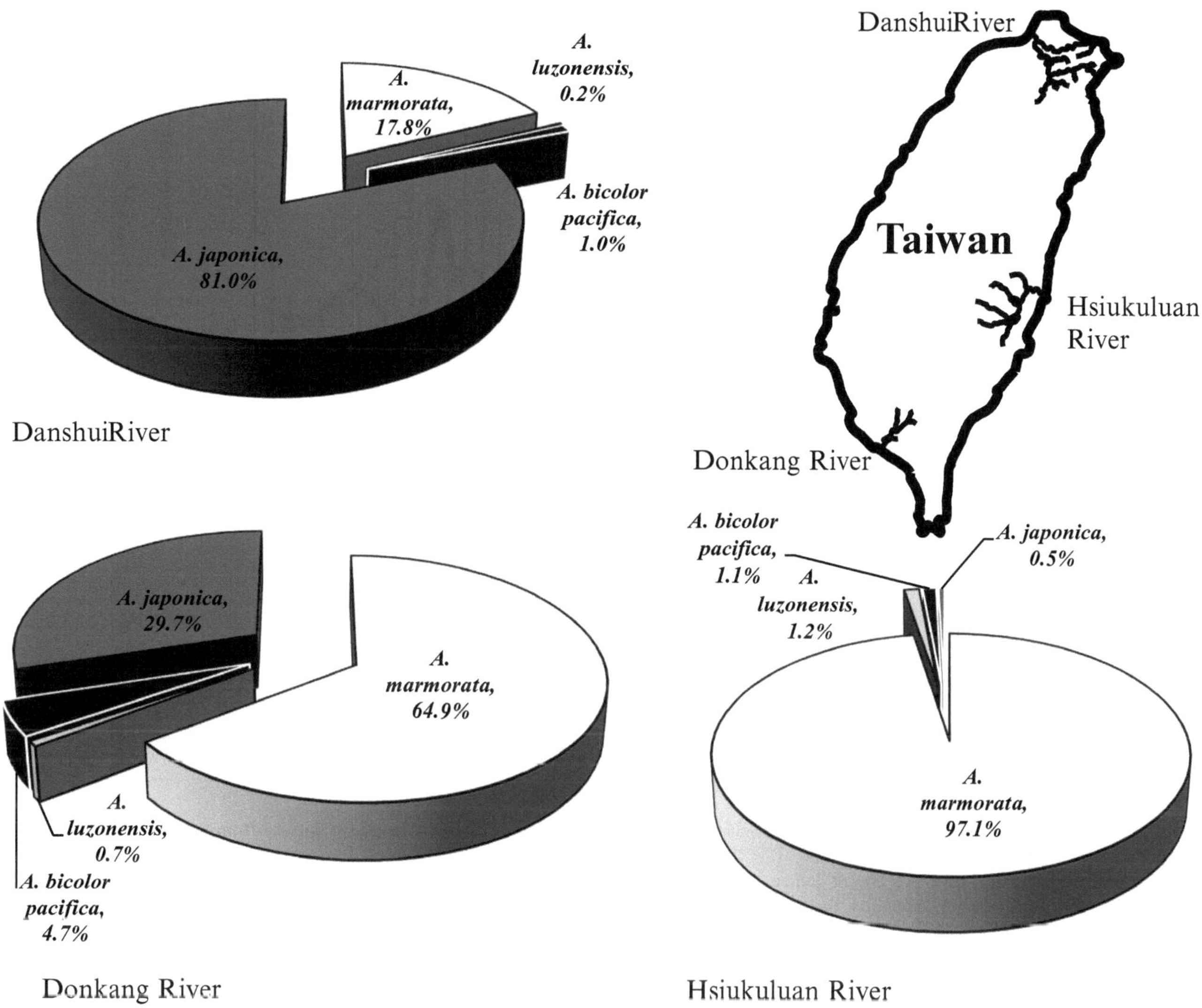

Fig. 9.2 Species compositions of glass eels caught in the Hsiukulan River estuary of eastern Taiwan (n = 3,035), the Tanshui River estuary of northwestern Taiwan (n = 2,583) and the Donkang River estuary of southwestern Taiwan (n = 148; Y-S. Han, unpublished data)

the two currents (NEC and Kuroshio) that transport eel larvae from the spawning grounds to the coasts and subsequently affects the Taiwanese glass eel catch is clearly a link (Tzeng et al. 2012a). After peaking in 1979, the Taiwanese glass eel catch gradually declined to a lower peak in 2001, since when it declined further until the most recent lower peak in 2011, mirroring similar decreases in the population size of the Japanese, American and European eels (Fig. 9.3b). All this is taken as evidence that fluctuations in the catches of glass eels in Taiwan reflect not only the overall population size of *A. japonica* but also ocean–atmosphere interactions exemplified by the climate change indices of sunspots and WPO.

Fishing for Eels in Taiwan

The types of fishery and fishing gear are diverse and reasonably unique to Taiwan. Both traditional and modern types of gear are deployed (Fig. 9.4). To catch inwardly migrating glass eels, *dipnets* with lights are used during flood tides at night by fishers standing on river banks or small fishing boats (Fig. 9.4a). Apparently, newly recruiting glass eels (Stages V_A and V_B) are relatively inactive and can be caught more easily than the pigmented stages (Stages VI_A or older) that have already adapted to an estuarine environment (Tzeng 1985). Also to catch glass eels, *handnets* are hauled through coastal waters near estuaries, held open by a stainless steel

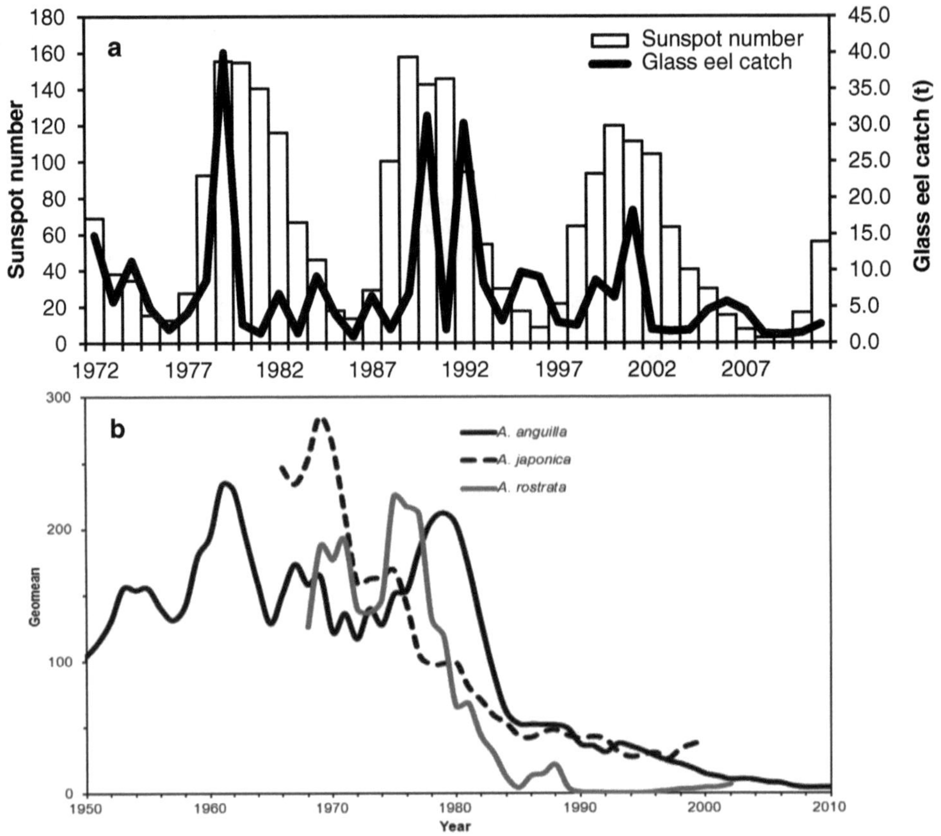

Fig. 9.3 (**a**) Annual trend in the Taiwanese catch of glass eels overlying a plot of the number of sunspots, 1972–2011. (**b**) Recruitment (3-year running averages of geomeans of indices as % 1979–1994 means) of glass European (*A. anguilla*), American (*A. rostrata*) and Japanese (*A. japonica*) eels (B. Knights, unpublished data). Note that the data for American eel are derived from yellow eels trapped at the Moses–Saunders Dam (St Lawrence River, Canada) and that the *plotted line* allows for the estimated 7-year lag between recruitment as glass eels and capture as yellow eels

frame (Fig. 9.4b) or by floaters and sinkers (Fig. 9.4c), and *triangular nets* (Fig. 9.4d) are operated by fishers pushing them against the current in estuaries and along nearby coasts. Finally, glass eels are captured by nets set across estuaries during flood tides at night (Fig. 9.4e), and because many such *setnets* are often laid out together at river mouths, boat passage can be impeded.

To catch adult eels, other types of gear are used. *Bamboo eel tubes* (Fig. 9.4f) have a long history of use, being designed before World War II and commonly used in the Nantou Prefecture, an area in central Taiwan well known for its bamboo production. The fishing gear is designed with a hole at one end of the bamboo into which the water flows to spread the smell of the bait (usually an earthworm) used to lure an eel into the funnel-shaped entrance at the other end. Recently, an alternative to bamboo in the form of a *PVC eel tube* has become more common; it is light and convenient to use, and one end of the tube is covered with a net while the other is left open. Unlike the bamboo tube, the *PVC* eel tube does not require bait, simply providing a ready-made shelter for the eel, and is set in both upper and lower reaches of rivers and in estuaries. Another means of catching adult eels uses the *shrimpnets* originally designed to catch river prawns but taking eels as a bycatch; eels are now targeted, however, and are caught in the nets when they enter in search of the prawns as food (Fig. 9.4g). The net is made of light nylon and contains several smaller units, each with a funnel-like opening that stops the eels inside from escaping. It is a highly effective means of catching adult eels and is moved around rivers easily by a small boat. Finally, *electrofishing* for eels in Taiwan used to be very popular (Fig. 9.4h) largely because it was a method that could be used irrespective of water currents, but it is now illegal in the country.

Eel Life History Determined from the Otolith

Although huge strides in knowledge have been made over the past few decades, the early life history of the Japanese eel while the leptocephali are being carried in the NEC and the Kuroshio Current from the spawning ground west of the Mariana Islands to estuaries along the coasts of East Asia is

Fig. 9.4 A montage of the various fishing gears used to catch eels in Taiwan: for glass eels in estuaries, (**a**) dipnet, (**b**) handnet with a stainless steel frame, (**c**) handnet with sinkers and floaters, (**d**) triangular net, (**e**) setnets; for yellow and silver eels in rivers, (**f**) bamboo eel tube (B, bottom; E, entrance), (**g**) shrimpnet, (**h**) electrofishing

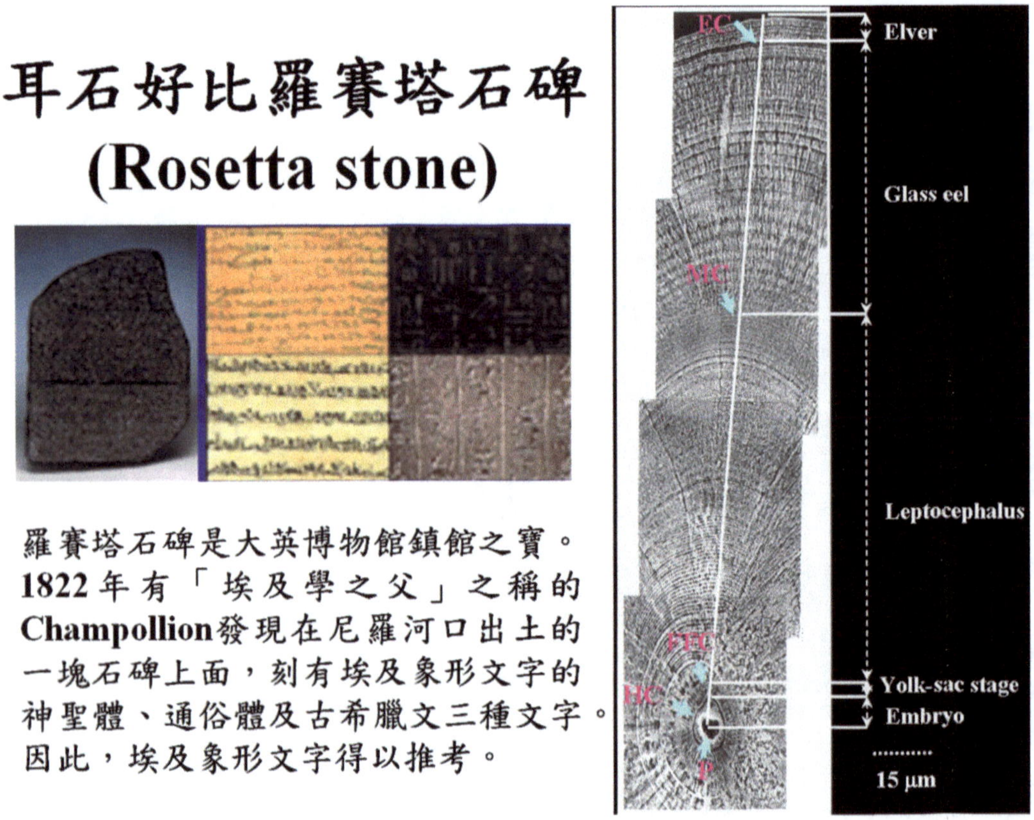

Fig. 9.5 Comparison of the archaeological records of ancient Egyptians displayed on the Rosetta stone (*left*) with records of the life history events of a young eel displayed in its otolith (*right*)

not yet fully understood. Some migratory life history information recorded in the otolith can be decoded through analyses of microstructure (Fig. 9.5). A fish otolith records life history events in a manner analogous to the Rosetta Stone (a well-known archaeological artefact currently housed at the British Museum in London, which has been used to unlock the mysteries of the hieroglyphs of ancient Egyptians).

An otolith is a biomineralized crystal of calcium carbonate ($CaCO_3$), functioning as an aid to hearing and balance in teleost (bony) fish, within which group eels are found. It acts as a time-keeper by depositing an annulus and daily growth increments (DGIs; Fig. 9.6) that allow one to determine the age of the fish in the form of annual zones and daily rings. A DGI is an unique structure, deposited in a circadian rhythm and allowing back-calculation of the birthdate and daily life of a fish by reading the rings (Tzeng 1990), so is regarded as a diary of the activities of a fish (Pannella 1971). An otolith is very like a CD-Rom, because the life history information of an eel recorded in its otolith can be retrieved in a manner similar to how one extracts data from a CD with a computer (Fig. 9.7). From DGIs identified in a scanning electron microscopic image of an otolith of a Japanese eel leptocephalus collected east of the Philippines (Fig. 9.8), it was discovered that the eel had been spawned during a new moon period (Liao et al. 1996), similar to that of the eel leptocephalus

Fig. 9.6 *Top:* Annuli and *Bottom:* daily growth increments (DGIs) in otoliths of a Japanese eel

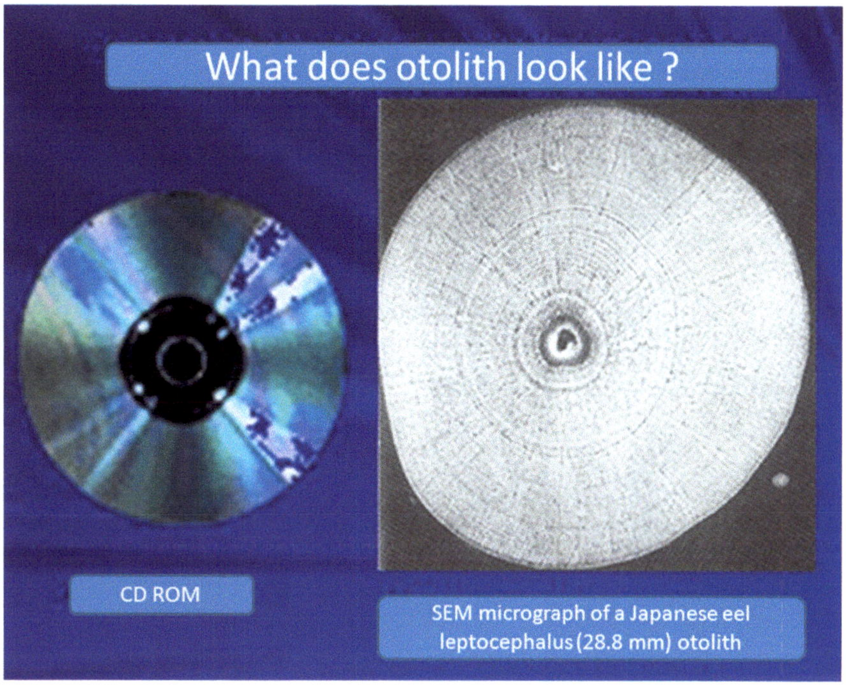

Fig. 9.7 A CD-Rom (*left*) and daily growth increments on a section of a sagittal otolith of a Japanese eel leptocephalus (*right*)

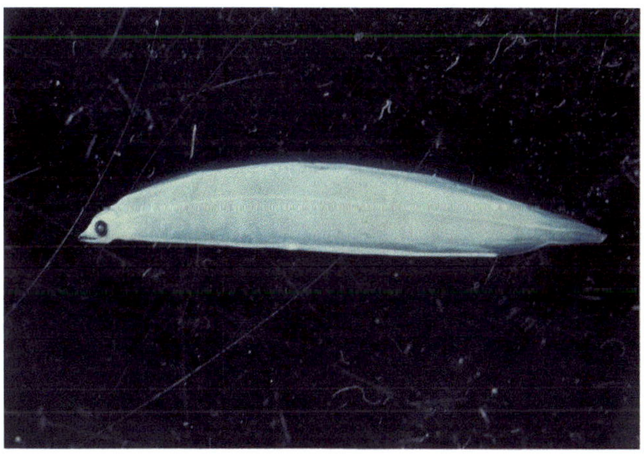

Fig. 9.8 At around 21:00 local time on 22 August 1995 at 12°30′N 131°30′E, Taiwanese researchers caught their first Japanese eel leptocephalus (27.40 mm total length) using an Isaacs–Kidd midwater trawl at a depth of 250 m

found at the Mariana Islands (Tsukamoto et al. 2003). It is from DGIs too that it was discovered that Japanese eel larvae take about 6 months to move passively with the NEC and Kuroshio Current over the ~5,000 km from the spawning grounds at the Mariana Islands to the estuaries of East Asia (Cheng and Tzeng 1996).

The otolith also records environmental information; its chemical composition is influenced by the ambient environment through which the fish passes during its migration or passive drift (Tzeng and Tsai 1994; Tzeng 1996). The otolith is inert, so once a trace element has been incorporated into its structure it will remain there for the life of the fish (Campana 1999), allowing one to reconstruct the past environmental history of an eel, for instance. That is the source of the recent discovery (mentioned above) that some Japanese eels spend their whole life in seawater; there are three types of residence determinable from temporal change in the Sr:Ca ratio in otoliths, seawater contingent, freshwater contingent and estuarine contingent (Fig. 9.9; Tzeng et al. 2002, 2003). In other words, part of the eel population can skip its freshwater phase (seawater contingent eels). Similar behaviour has since been found for the European eel (Tzeng et al. 1997, 2000; Daverat et al. 2006; Shiao et al. 2006) and the American eel (Jessop et al. 2006; Thibault et al. 2007; Lamson et al. 2009), and it is therefore likely that such facultative migratory behaviour is advantageous to a fish adapting over time to a changing natural environment.

Evolution of the Eel Aquaculture Industry in Taiwan

The development of Taiwan's aquaculture industry for *A. japonica* can be divided into seven stages (Fig. 9.10; Hsue 2012). Most culture is carried out outdoors (Fig. 9.11). It started in 1952, initially at a small scale, but from 1970, production increased rapidly and peaked at about 55,000 t with a value of ~NT$15 billion (equivalent to US$518 million at the 2012 exchange rate) in the early 1990s. Of the total production of Japanese eels in Taiwan then, ~90 % was exported to Japan and the other 10 % consumed locally. However, annual production in Taiwan then dropped dramatically, to <20,000 t

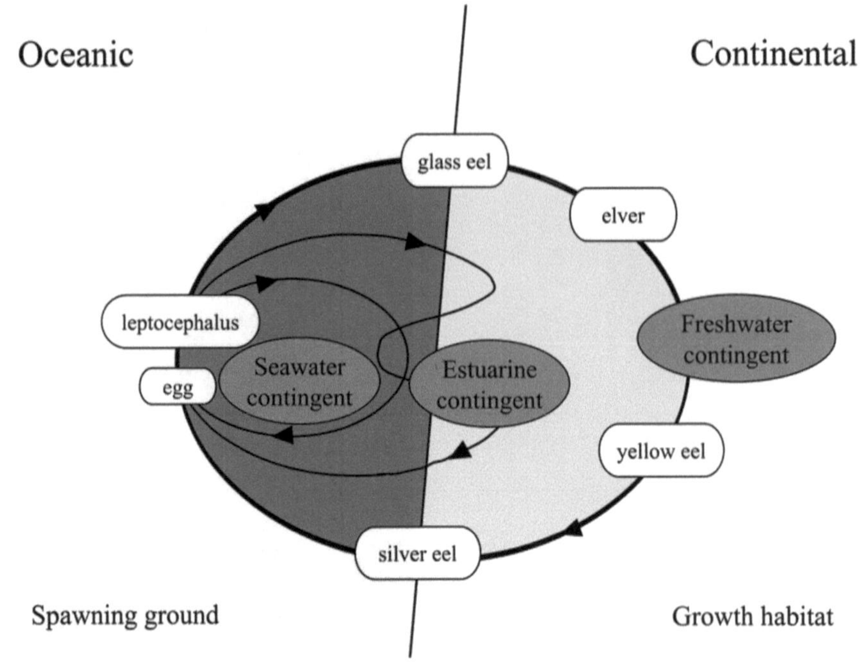

Fig. 9.9 The life history of freshwater eels displayed showing the three migratory patterns possible, freshwater contingent, estuarine contingent and seawater contingent

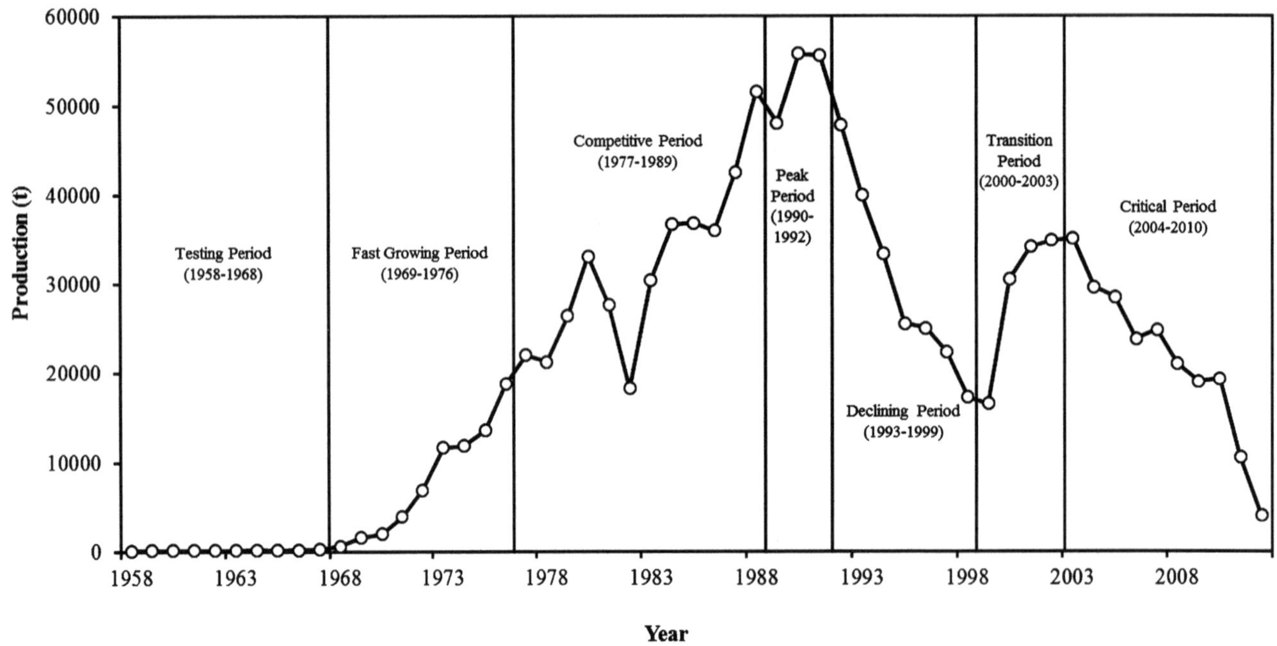

Fig. 9.10 Aquaculture production of Japanese eels in Taiwan, 1958–2012, showing the seven phases mentioned in text (updated from Hsue 2012)

by 1999, almost certainly influenced by the reduction in imports of glass eels from China that resulted from the rise of eel aquaculture there for export to Japan following China's economic revolution in 1980. The annual aquaculture production of eels in mainland China was ~165,000 t in 1999, of which ~95,000 t was exported to Japan, so mainland China over time clearly replaced Taiwan as the source of material destined for consumption in Japan. With land and labour costs cheaper in China than in Taiwan, it was obviously economically beneficial for Taiwanese eel farmers to relocate their businesses to mainland China. Taiwanese eel production did rise again slightly, concomitant with increasing catches of glass eels, between 2000 and 2003, but it thereafter decreased again to an annual value of <10,000 t (Fig. 9.10).

Fig. 9.11 Outdoor aquaculture of Japanese eels in Taiwan: (**a**) a water wheel used to increase the dissolved oxygen; (**b**) artificial feed being provided in a cage; (**c**) eels competing for food

On average, China and Japan account for ~80 % of the glass eel catches by Asian countries and Taiwan and Korea the remaining ~20 %. The annual catches of glass eels by all Asian countries combined have varied over the years, but Taiwan's catch peaked in the 1970s before gradually decreasing to its very low level today (Fig. 9.3a). Some 50–75 % of the glass eels taken in Taiwan are harvested before upstream migration (Tzeng 1984), but the catch has traditionally been much less than the demand, so imports from other countries have been part of the eel aquaculture industry since it burgeoned in the 1960s and 1970s. Overall, the relatively limited local catch and increased competition among producers of glass eels for the increasingly expensive (local and imported) raw material has been behind the general reduction in eel aquaculture in Taiwan over the past two decades. Eel aquaculture contributed greatly not only to Taiwan's national economic development, but also notably to the development of local fishing villages. However, it did have some negative environmental consequences: the overextraction of underground freshwater for eel aquaculture caused widespread land subsidence, and the use of seawater for aquaculture raised the salinity of the soil in some adjacent rice fields. Such adverse effects have been less evident since the late 1990s, though, the area under aquaculture now being less, by recycling water, and by the use of surplus ponds for rainwater collection and storage.

Taiwanese Conservation and Restocking Programmes

In an attempt to supplement the production of Japanese eel eggs and larvae in the wild and thence hopefully to stimulate the recruitment of glass eels to Taiwanese estuaries for aquaculture as well as to promote a sustainable fishery for adult eels, the Taiwan Fisheries Research Institute (TFRI) released hormone-induced eel spawners into the Shiau-Liu-Chio Trench (120°20′–36′E 22°15′–30′N) off the country's southwestern coast. Between 1976 and 2001, a total of 29,921 adult eels weighing >50 t was released during 23 cruises made to the trench between October and March each year (mostly during December). The trench was thought at that time to be the spawning ground for the Taiwanese population of Japanese eels (Kuo 1971; Kimura et al. 1999). However, the Japanese eel is now known to be a panmictic population (Han et al. 2010), and its only known spawning ground is in waters west of the Mariana Islands (142°E 14°N; Tsukamoto 1992). It is unknown whether those releases of adult eels had any positive effect on the population, and in May/June of 1999–2001, three additional TFRI cruises were made to the now-known spawning site, releasing another 1,500 hormone-induced spawners weighing >2 t. To date, it has proven impossible to evaluate the effect of this release programme on the recruitment of Japanese glass eels to Taiwan, and the fate of individual eels remains unknown. Certainly, however, there is no significant relationship between the annual numbers of spawners released at sea between 1976 and 2001 and the annual catches of glass eels in Taiwan (Tzeng et al. 2012b).

In the early 2000s, following criticism by the international scientific community that such a release programme in the open ocean might induce genetic pollution, the programme was stopped, and since 2003, the TFRI has restocked freshwater streams only. During the months July–November of 2003–2008, 35,636 young eels weighing >76 t were released into three of Taiwan's main rivers, one each in the northwest, the northeast and the southwest. These young eels have been relatively easy to monitor, and mark–recapture data have been collected to evaluate the effects of the restocking programme on exploitation status and to collect information and parameters important in furthering understanding of the species' population dynamics and migratory behaviour. Following analysis using a standard yield-per-recruit model, it was evident that growth-overfishing was taking place in Kao-Ping River in southwestern Taiwan, i.e. too many small eels were being caught before optimal growth had been attained. A spawner-per-recruit model demonstrated recruit-overfishing too, i.e. insufficient silver eels escaping from the river back to sea to spawn (Lin et al. 2009, 2010; Lin and Tzeng 2010; Tzeng 2012).

Eel Cuisine in Taiwan

More than half of Taiwan's production of Japanese eels is converted into "kabayaki" and "shirayaki" for export to Japan (Fig. 9.12a, b). Japanese people generally celebrate "Ushinohi" in July by consuming eels. In Taiwan, however, dishes of eel meat, which are high in energy because of the fat content, are enjoyed by locals throughout winter, particularly on the day of the winter solstice (21 or 22 December), to keep the body warm; although the country is classified as subtropical, Taiwanese winters can be very cold.

To accommodate local consumption, a diversity of eel dishes has developed over the years. Taiwanese people generally prefer the giant mottled eel over the Japanese eel as food, but the former species was listed as endangered from 1989 to 2009 and its culture, sale and fishing banned. Recently, however, the ban has been lifted and the giant mottled eel has again become a popular food and is listed on the menus of many of the popular restaurants on the island (Fig. 9.12c–f).

Arts and Mythology

The eel is an important religious icon in Taiwanese folklore. The Japanese eel and the giant mottled eel are regarded, respectively, as river and sea gods, and this can be seen in the design and paintings of gate god statues commonly placed at the entrance to traditional Taiwanese village dwellings. Many villagers believe that the gate gods protect them against the devil and evil spirits, and protect the security of their family (Fig. 9.13a). Additionally, eels appear in the design of "ong-bao" (Fig. 9.13b), the red bags containing money that parents give children to seek good fortune during the Chinese Lunar New Year. Eels are important also for Taiwanese native (aboriginal) peoples, but those people do not kill and eat the eels because they believe that they are the embodiment of celestial beings.

The eel is also considered to be a propitious creature in Taiwan (and elsewhere), and this can be seen from its depiction on things used commonly in daily life, such as necktie pins, company logos and souvenirs (Fig. 9.14).

The Future of Eels in Taiwan

Both the fishery and aquaculture of Japanese eels in Taiwan have a long history, contributing greatly to local economic development, and eel dishes have been developed to satisfy the needs of people in their daily lives. However, the natural population of Japanese eels has decreased dramatically over the past three decades, similar to the situation with the American eel and the European eel, and the catch of glass eels has decreased to a level much lower than that needed to meet aquaculture demand. To sustain eel aquaculture into the future, therefore, it is essential that enhanced effort be put into developing the means of artificially propagating the seed eels for aquaculture (see the Japanese and Danish chapters for

Fig. 9.12 Some examples of cuisine offered in Taiwanese restaurants: (**a**) kabayaki, (**b**) shirayaki, (**c**) fried eel, (**d**) eel prepared in a casserole with ginger, basil and wine, (**e**) eel boiled with Chinese herbal medicine, (**f**) shabu-shabu (**a** and **b**, Japanese eel; **c–f**, Giant mottled eel). Photographs reproduced with permission (**a** and **b**) Just Champion Enterprise Co. Ltd, (**c–e**) Ji-Peng Chen/Super Utmost, (**f**) Jun-Ren Chen

information on progress to date), to preclude overfishing glass eels in east Asia and elsewhere. In addition, it is crucial that the fisheries for yellow and silver eels in rivers be better regulated and that anthropogenic effects on the habitats of eels be minimized if any recovery of wild eel stocks is to materialize in future.

Finally, although it is abundantly clear that eels are hugely important to the people of Taiwan, as they are in much of the rest of East Asia, serious collaborative research and development efforts and definitive management are needed if eel populations are to be preserved for the benefit of future generations of humankind.

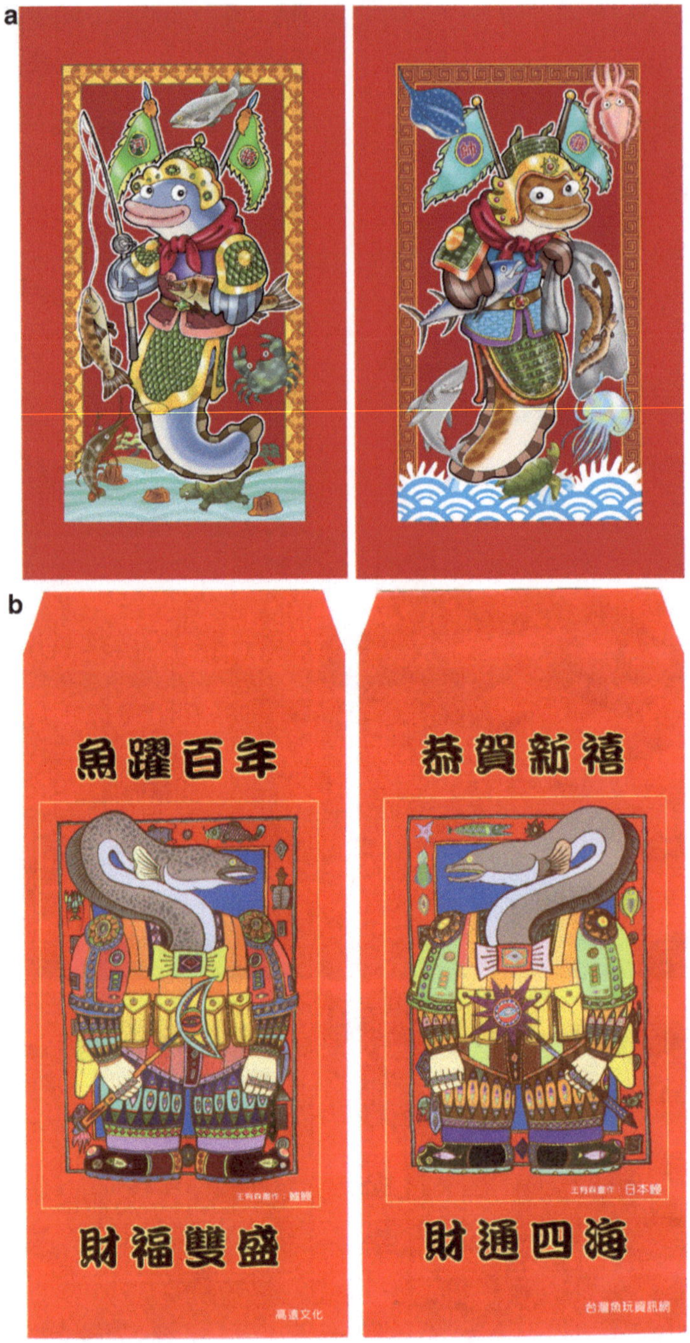

Fig. 9.13 (**a**) Gate gods depicting *A. japonica* (*left*) and *A. marmorata* (*right*), and (**b**) red bags with A. *marmorata* (*left*) and *A. japonica* (*right*) shown as symbols of good fortune for Chinese New Year celebrations. Reproduced with permission (**a**) De-Hong Shih/Super Utmost and (**b**) You-Sen Wang/Super Utmost

Fig. 9.14 A gold-plated necktie pin depicting a Japanese eel (*top left*), the logo of an eel marketing company in Northern Ireland (*bottom left*), and a commemorative mug with an eel design presented to the author on his retirement (*right*)

Acknowledgements I thank the National Science Council of the Republic of China for long-term financial support for eel research, Brian Knights and Yu San Han for providing unpublished data, Nico Jose Leander for producing much of the artwork, and Brian M. Jessop, Chu-Fa Tsai and book series editor Andrew I. L. Payne for comments, including grammatical improvements, on various earlier drafts.

References

Campana SE (1999) Chemistry and composition of fish otoliths: pathways, mechanisms and applications. Mar Ecol Prog Ser 188: 263–297

Chang S-T, Yeh M-F, Chen R-T, Tsai C-L, Shen K-N, Tsai C-F (2008) A giant exotic speckled longfin eel (*Anguilla reinhardtii* Steindachner, 1867) captured from the Sun Moon Lake. Endem Species Res (Taipei) 10(2):59–64

Chang K-C, Huang Y-S, Liu F-K (2012) How many *Anguilla* species in the Taiwan waters? In: Proceedings of the 15th East Asia eel resource consortium, National Taiwan Ocean University, Keelung, Taiwan, p 35

Cheng P-W, Tzeng W-N (1996) Timing of metamorphosis and estuarine arrival across the dispersal range of the Japanese eel *Anguilla japonica*. Mar Ecol Prog Ser 131:87–96

Daverat F, Limburg KE, Thibault I, Shiao J-C, Dodson JJ, Caron F, Tzeng W-N et al (2006) Phenotypic plasticity of habitat use by three temperate eel species, *Anguilla anguilla*, *A. japonica* and *A. rostrata*. Mar Ecol Prog Ser 308:231–241

Han Y-S, Chang C-W, He I-T, Tzeng W-N (2001) Validation of the short-finned eel *Anguilla bicolor pacifica* in natural waters of Taiwan. Acta Zool Taiwanica 12:9–19

Han Y-S, Yu C-H, Yu H-T, Chang C-W, Liao I-C, Tzeng W-N (2002) The exotic American eel in Taiwan: ecological implications. J Fish Biol 60:1608–1612

Han Y-S, Hung C-L, Liao Y-F, Tzeng W-N (2010) Population genetic structure of the Japanese eel *Anguilla japonica*: panmixia at spatial and temporal scales. Mar Ecol Prog Ser 401:221–232

Han Y-S, Zhang H, Tseng Y-H, Shen M-L (2012) Larval Japanese eel (*Anguilla japonica*) as sub-surface current bio-tracers in East Asia continental shelf. Fish Oceanogr 21:281–290

Hsue W-C (2012) Cost and benefit analysis of *Anguilla japonica* and *A. marmorata* aquaculture in Taiwan. Master thesis, Institute of Fisheries Science, College of Life Science, National Taiwan University

Jessop BM, Shiao J-C, Iizuka Y, Tzeng W-N (2006) Migration of juvenile American eels *Anguilla rostrata* between freshwater and estuary, as revealed by otolith microchemistry. Mar Ecol Prog Ser 310:219–233

Kimura S, Döös K, Coward AC (1999) Numerical simulation to resolve the issue of downstream migration of the Japanese eel. Mar Ecol Prog Ser 186:303–306

Kuo H (1971) Approach of eel elvers to the land in Taiwan. Fish Cult 8:52–56 (in Japanese)

Lamson HM, Cairns DK, Shiao J-C, Iizuka Y, Tzeng W-N (2009) American eel, *Anguilla rostrata*, growth in fresh and salt water:

implications for conservation and aquaculture. Fish Manag Ecol 16:306–314

Leander NJ, Shen K-N, Chen R-T, Tzeng W-N (2012) Species composition and seasonal occurrence of recruiting glass eels (*Anguilla* spp.) in the Hsiukuluan River, eastern Taiwan. Zool Stud 51: 59–71

Liao I-C, Kuo C-L, Tzeng W-N, Hwang S-T, Wu C-L, Wang C-H, Wang Y-T (1996) The first time of leptocephali of Japanese eel *Anguilla japonica* collected by Taiwanese researchers. J Taiwan Fish Res 4(2):107–116

Lin Y-J, Tzeng W-N (2010) Vital population statistics based on length frequency analysis of the exploited Japanese eel (*Anguilla japonica*) stock in the Kao-Ping River, southern Taiwan. J Appl Ichthyol 26:424–431

Lin S-H, Chang S-L, Iizuka Y, Chen T-I, Liu F-G, Su M-S, Su W-C (2009) Use of mark-recapture and otolith microchemistry on the study of the migratory behavior and habitat use of Japanese eels (*Anguilla japonica*). J Taiwan Fish Res 17(2):47–65

Lin Y-J, Chang Y-J, Sun C-L, Tzeng W-N (2010) Evaluation of the Japanese eel fishery in the lower reaches of the Kao-Ping River, southwestern Taiwan using a per-recruit analysis. Fish Res 106:329–336

Pannella G (1971) Fish otoliths: daily growth layers and periodical patterns. Science 173:1124–1127

Shiao J-C, Iizuka Y, Chang C-W, Tzeng W-N (2003) Disparities in habitat use and migratory behavior between tropical eel *Anguilla marmorata* and temperate eel *A. japonica* in four Taiwanese rivers. Mar Ecol Prog Ser 261:233–242

Shiao J-C, Ložys L, Iizuka Y, Tzeng W-N (2006) Migratory patterns and contribution of stocking to the population of European eel in Lithuanian waters as indicated by otolith Sr:Ca ratios. J Fish Biol 69:749–769

Teng H-Y, Lin Y-S, Tzeng C-S (2009) A new *Anguilla* species and a reanalysis of the phylogeny of freshwater eels. Zool Stud 48: 808–822

Tesch F-W (2003) The eel, 5th edn. Blackwell Publishing, Oxford, 408 pp

Thibault I, Dodson JJ, Caron F, Tzeng W-N, Iizuka Y, Shiao J-C (2007) Facultative catadromy in American eels: testing the conditional strategy hypothesis. Mar Ecol Prog Ser 344:219–229

Tsukamoto K (1992) Discovery of the spawning area for Japanese eel. Nature 356:789–791

Tsukamoto K, Arai T (2001) Facultative catadromy of the eel *Anguilla japonica* between freshwater and seawater habitats. Mar Ecol Prog Ser 220:265–276

Tsukamoto K, Otake T, Mochioka N, Lee T-W, Fricke H, Inagaki T, Aoyama J et al (2003) Seamounts, new moon and eel spawning: the search for the spawning site of the Japanese eel. Environ Biol Fishes 66:221–229

Tzeng W-N (1982) Newly record of the elver, *Anguilla celebesensis* Kaup, from Taiwan. Chin Biosci (Taipei) 19:57–66 (in Chinese with English abstract)

Tzeng W-N (1983) Species identification and commercial catch of the anguillid elvers from Taiwan. China Fish Mon 366:16–23 (in Chinese with English abstract)

Tzeng W-N (1984) An estimate of the exploitation rate of *Anguilla japonica* elvers immigrating into the coastal waters off Shuang-Chi River, Taiwan. Bull Inst Zool, Academia Sinica 23:173–180

Tzeng W-N (1985) Immigration timing and activity rhythms of the eel, *Anguilla japonica*, elvers in the estuary of northern Taiwan, with emphasis on environmental influences. Bull Jap Soc Fish Oceanogr 47(48):11–28

Tzeng W-N (1990) Relationship between growth rate and age at recruitment of *Anguilla japonica* elvers in a Taiwan estuary as inferred from otolith growth increments. Mar Biol 107:75–81

Tzeng W-N (1996) Effects of salinity and ontogenetic movements on strontium: calcium ratios in the otoliths of the Japanese eel, *Anguilla japonica* Temminck and Schlegel. J Exp Mar Biol Ecol 199:111–122

Tzeng W-N (2012) A review of the population structure, life history strategy, and sustainable fishery of the Japanese eel *Anguilla japonica* in Taiwan. J Fish Soc Taiwan 38(1):1–19

Tzeng W-N, Tabeta O (1983) First record of the short-finned eel *Anguilla bicolor pacifica* elvers from Taiwan. Bull Jap Soc Scientific Fish 49:27–32

Tzeng W-N, Tsai Y-C (1994) Changes in otolith microchemistry of the Japanese eel, *Anguilla japonica*, during its migration from the ocean to the rivers of Taiwan. J Fish Biol 45:671–683

Tzeng W-N, Cheng P-W, Lin F-Y (1995) Relative abundance, sex ratio and population structure of the Japanese eel *Anguilla japonica* in the Tanshui River system of northern Taiwan. J Fish Biol 46:183–200

Tzeng WN, Severin KP, Wickström H (1997) Use of otolith microchemistry to investigate the environmental history of European eel *Anguilla anguilla*. Mar Ecol Prog Ser 149:73–81

Tzeng WN, Wang CH, Wickström H, Reizenstein M (2000) Occurrence of the semi-catadromous European eel *Anguilla anguilla* (L.) in the Baltic Sea. Mar Biol 137:93–98

Tzeng WN, Shiao JC, Iizuka Y (2002) Use of otolith Sr:Ca ratios to study the riverine migratory behaviours of Japanese eel *Anguilla japonica*. Mar Ecol Prog Ser 245:213–221

Tzeng W-N, Iizuka Y, Shiao J-C, Yamada Y, Oka HP (2003) Identification and growth rates comparison of divergent migratory contingents of Japanese eel (*Anguilla japonica*). Aquaculture 216:77–86

Tzeng W-N, Han Y-S, Jessop BM (2009) Growth and habitat residence history of migratory silver American eel transplanted to Taiwan. Am Fish Soc Symp 58:137–147

Tzeng W-N, Han Y-S, Tsukamoto K, Kuroki M (eds) (2012a) Story of eel. Lanyang Museum, Taiwan, 232 pp (in Chinese)

Tzeng W-N, Tseng Y-H, Han Y-S, Hsu C-C, Chang C-W, Di Lorenzo E, Hsieh C-H (2012b) Evaluation of multi-scale climate effects on annual recruitment levels of the Japanese eel, *Anguilla japonica*, to Taiwan. PLoS One 7(2):e30805. doi:10.1371/journal.pone.0030805

Watanabe S, Aoyama J, Tsukamoto K (2009) A new species of freshwater eel *Anguilla luzonensis* (Teleostei: Anguillidae) from Luzon Island of the Philippines. Fish Sci 75:387–392

Williamson GR, Boëtius J (1993) The eels *Anguilla marmorata* and *A. japonica* in the Pearl River, China, and Hong Kong. Asian Fish Sci 6:129–138

Freshwater Eels and People in New Zealand: A Love/Hate Relationship

10

Don Jellyman

New Zealand has a relatively small native fish fauna, just 38 described species (McDowall 1990), a small number relative to the 95 in Japan, a land of similar size. Eels are the largest and most frequently encountered native species in New Zealand, and often make up 90 % or more of the biomass (total weight) of fish in a river or stream (Rowe et al. 1999).

There are three species of freshwater eel in New Zealand (Fig. 10.1), two shared with Australia (the shortfin eel, *Anguilla australis*, and the Australian spotted longfin eel *A. reinhardtii*). The third species, the New Zealand longfin eel (*A. dieffenbachii*) is found only in New Zealand and its offshore islands. The mottled (Australian) longfin is uncommon and, being a tropical species, is found in New Zealand only in northern North Island. The other two species are widespread throughout the country, shortfin eels mainly in slower, muddier coastal waters, and longfin eels in clearer, faster-flowing rivers and high-country lakes. The shortfin is seldom >1 m long (and 3–4 kg), but the longfin grows much larger, sometimes exceeding 1.5 m and 20+ kg, and occasionally as long as 2 m (50 kg). As with freshwater eels worldwide, the larger individuals are females.

Eel Biology

The baby eels, the glass eels (5.5–6.5 cm long), arrive in New Zealand streams and rivers in spring, mainly in September and October. When they first arrive in freshwater they are active only at night, but as they adapt to their new habitat, they commence forming shoals and often migrate upstream during daylight. Then, in summer each year for several years thereafter, the small eels, now known as elvers, migrate upstream, and in that way the longfin eels especially find their way to the very upstream reaches of rivers. Having found a suitable place to live, the eels often stay in the same location for many years. They feed opportunistically on a wide range of food, depending on availability, and during flooding, many feed in flooded pastures on earthworms, grubs and spiders. Larger eels of both main species eat small fish when they are available; this high energy diet results in more-rapid growth.

Growth of eels in New Zealand is usually slow, some 2–4 cm year^{-1} (Jellyman 2009), so a large longfin eel can be 50 or more years old when eventually migrating back to the sea to spawn. Indeed, in areas where productivity is low, longfins grow even more slowly than that and have been aged at >100 years (Jellyman 1995). The spawning migration downstream takes place in distinct seasonal phases, with shortfins preceding longfins, and males of both species preceding females. The spawning grounds of New Zealand eels are not known with certainty, but the results from tracking female longfin eels equipped with satellite tags (Jellyman and Tsukamoto 2010) and from models of possible larval dispersal using surface currents measured from satellites (Jellyman and Bowen 2009) have both indicated that it is probably in the South Fiji basin.

Eels and Māori

Before the arrival of Europeans in New Zealand a little over 200 years ago, Māori (the indigenous people of New Zealand) had an excellent understanding of the biology of eels along with the timing and conditions of when they migrated, and they had developed very effective means of catching them. Because they were an important source of food, were plentiful and grew to such a large size, eels feature extensively in Māori legends, proverbs and stories.

The Names of Eels

The most common Māori name for eels is *tuna*. As this is the same name as the marine species that includes bluefin tuna,

D. Jellyman (✉)
National Institute of Water and Atmosphere,
8602, Christchurch 8440, New Zealand
e-mail: d.jellyman@niwa.co.nz

K. Tsukamoto and M. Kuroki (eds.), *Eels and Humans*, Humanity and the Sea,
DOI 10.1007/978-4-431-54529-3_10, © Springer Japan 2014

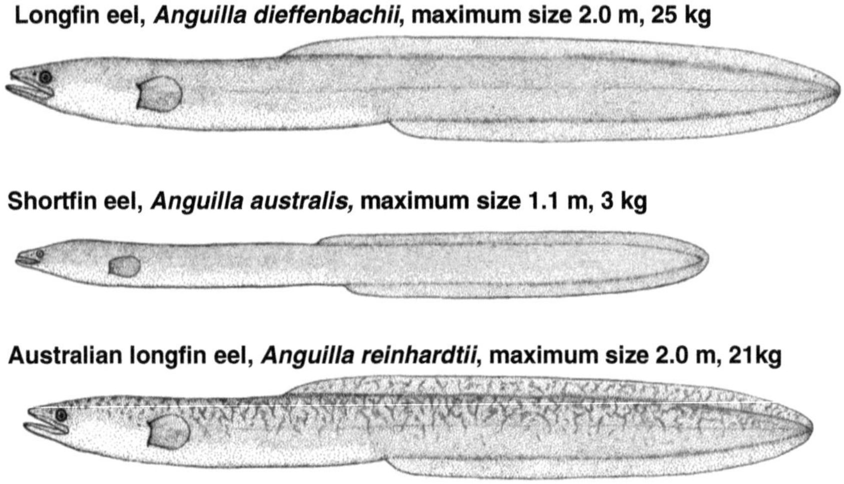

Fig. 10.1 New Zealand freshwater eels. Drawings courtesy R. M. McDowall

bonito and skipjack, the name does sometimes cause confusion with overseas researchers who unsurprisingly do not understand stories of Māori catching tuna in freshwater using earthworms as bait! Māori had many different names for *tuna* (eels), depending on the Māori tribe, but also on species, size, colour, life stage, season, etc; well over 140 separate names have been recorded to date (Best 1929). *Heke* is the Māori word describing the annual eel migration to sea, so a *tuna-heke* is a migrating (silver) eel.

Certain tribes were renowned for the quality and flavour of their eels, and eels would frequently be exchanged as gifts (*koha*) when visiting a neighbouring tribe. If a tribe renowned for its eels could not provide at least some eels for food, gifts or barter, it could result in a significant loss of prestige (*mana*). These traditions are still carried on today, but being able to supply enough eels of suitable size tends to be difficult and has led to many complaints to fishery managers about overfishing by commercial eel fishers.

Importance of Eels

Eels were especially important to Māori because there were no native mammals in New Zealand (apart from two small bats) and forest birds were often small and not abundant. In contrast, eels were plentiful, distributed widely and easily caught; they were also by far the largest of the native fish. More than any other food too, eels provided fat and oil in Māori diet. Further, they could be retained alive until needed in specially woven containers (corfs); sometimes the eels so retained were fed on food such as kumara (sweet potato). Once killed, they could be eaten fresh or preserved by smoking or by drying in the sun. Preserved eels could be kept for months in bags made of kelp or bark, and whale oil was sometimes poured over the contents to exclude the air and to preserve the eels. Preservation was very important because over much of New Zealand, eels were only available seasonally—cool water temperatures during winter cause eels to become inactive, meaning that they cannot be caught, and Māori relied upon preserved eels for food during the colder months. The numbers of eels eaten by Māori were often large, and one historian recalled a feast where 20,000 dried eels were available. The value of eels was such that one European historian claimed that eel fisheries were "as valuable to the ancient Māoris as gold-mines are to Europeans" (McDowall 2011).

Fishing Methods and Seasons

Māori developed an excellent and detailed understanding of the biology of eels, especially the seasonal movements that eels undergo. There are no records of Māori harvesting glass eels when they enter freshwater from the sea, but elvers were often caught in large quantities when they congregated below waterfalls. The usual method was to place bundles of ferns and twigs near the base of the falls, and every few days shaking the small eels out. At some places, it was not unusual for 500–750 kg to be harvested in this way in a single night. Scientists have since used the same method to collect samples of small eels in lakes.

Eel fishing was often accompanied by religious rites and ceremonies, usually preceded by a blessing (*karakia*) from a priest (*tohunga*) to ensure the success of the fishing and the safety of the fishers. Carved ceremonial stakes were placed at strategic areas to designate ownership of an eel weir or fishing rights to an area of river, and poaching was punishable by death.

The larger eels were generally caught by spearing or bobbing. Multiheaded spears made from hardwood (Fig. 10.2) were used at night with the aid of a burning torch. Bobbing required the fisher to gather earthworms and grubs and to thread them carefully lengthwise onto a piece of flax or thin tree bark; sometimes "glow-worms" (worms that have a luminescent glow) were used. Several of these strands were then gathered together, sometimes laced around with further flax strands. The bob was then tied to a stout stick and dropped into the water in a promising fishing spot, close to a place where eels were likely to be living. The eels were caught when their small, backward-facing teeth became tangled in the flax fibres.

Eels were also caught by hand by fishers probing under banks and in holes (Fig. 10.3). Usually the fisher's hand slid along the eel's body until it could be clamped behind the head, or fingers inserted into the gill openings; the eels so caught then had to be thrown ashore quickly, to be clubbed. Fishers would also wade across muddy and shallow areas, feeling for eels with their feet. In shallow lakes subject to strong winds, eels were also caught by hand when the wind blew the water away from shallow parts of the lake bed.

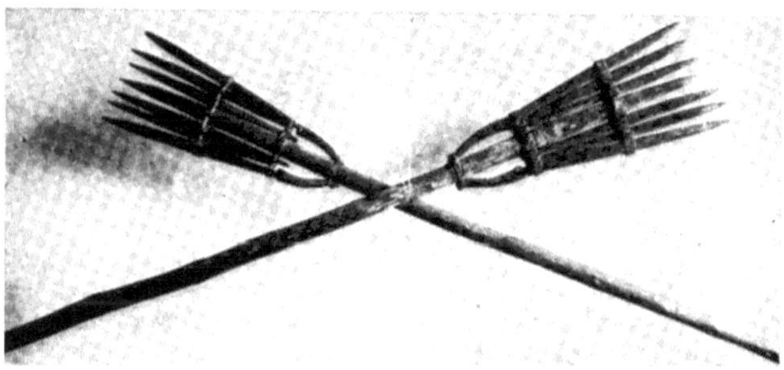

Fig. 10.2 Māori eel spears made of wood (photo courtesy Museum of New Zealand, Te Papa Tongarewa)

In many lowland areas and estuaries, canals were dug to encourage eels to move along certain routes where they could be more easily caught. These canals, all dug by hand with wooden digging sticks, were sometimes extensive, with one lagoon area having >20 km of canals (McDowall 2011).

Clubs were used by Māori to stun eels in shallow water, and spears were also used, especially when migrating eels could be encouraged to enter shallow channels dug into the banks separating some lakes and lagoons from the sea. With the arrival of European settlers who brought steel, gaffs became a popular means of harvesting eels. Gaffing of silver eels entering dug channels is still practiced today in a lake reserved for Māori eel-fishing near Christchurch; there, the migrating (*heke*) eels swim into hand-dug trenches, especially on dark nights, and are gaffed and flung into shallow pits in the gravel (Fig. 10.4). When activity has finished for the night, the eels are gathered in sacks, killed by adding salt to them, gutted and their backbone removed, partly air-dried, then smoked for preservation purposes. Although there are more modern and efficient means available today for catching eels, the local tribe prefers to maintain its old tradition of eel capture and preservation (Fig. 10.5).

The most widespread method of capture by Māori, however, was to use fish traps (*hīnaki*). These were traps made from plaiting flax, durable vines and bark (Fig. 10.6), with a funnel entrance one end that generally tapered to a removable cover at the other end from which the catch could be tipped out. They were often baited with birds or worms (and occasionally with human flesh!), and set overnight,

Fig. 10.3 An old drawing showing Māori catching eels by hand (from New Zealand Graphic, 29 December 1894, reproduced with permission)

Fig. 10.4 Wairewa (Lake Forsyth): (**a**) trenches dug in the outlet by hand that migrating eels enter at night; (**b**) migrating eels entering the trench; (**c**) migrating eels being gaffed from the trench (**b** and **c** courtesy Jack Jacobs, Wairewa runanga)

Fig. 10.5 A drying rack covered with migrating shortfin eels at Wairewa (Lake Forsyth, South Island, New Zealand) (**a**) in 1948 (photo courtesy Alexander Turnbull Library), and (**b**) one in use today (photo courtesy Leah Brown)

sometimes at the top of a small fence which served to direct eels towards the entrance of the trap.

The most engineeringly complex method of capture by Māori, however, used *pā-tuna* (large weirs; Fig. 10.7), which trapped large numbers of migrating (silver) eels in autumn and early winter. These were impressive structures, sometimes built completely across a river, or at an angle to the main flow (Downes 1918). They were designed mainly to entrap migrating eels, and the fences were used to direct downstream migrating eels into traps (*hīnaki*). During a visit to New Zealand, Danish eel biologist Johannes Schmidt, renowned for solving the mystery of where European and American eels spawned (see the Danish chapter in this book), observed that Māori "methods of capture had attained a high degree of technique .... These weirs in particular are—or were—in structure, extent and position, sometimes in connection with artificial canals, calculated to arouse the admiration of any fisheries expert..." (Schmidt 1925). Some weirs were so large as to obstruct river transport, such as paddle streamers, so were subsequently removed by European settlers. Weirs were widespread and common; in one lower tributary of the Whanganui River (a major North Island river), for instance, there were once 56 Māori-constructed weirs. Such was their importance that ownership of weirs was handed down to succeeding generations within the same family.

Fig. 10.6 Large woven traps *(hinakī)* used to catch eels (after Best 1929) (photo courtesy Museum of New Zealand, Te Papa Tongarewa)

Weirs were built by hand, small ones sometimes from stones, and larger ones always from timber. The main timber posts were usually 9–15 cm in diameter and were driven into the riverbed with large wooden hammers (mauls). Mats were laid where the riverbed might become scoured. Emptying or changing the *hīnaki* was a risky operation, because most eel migrations took place when rivers ran high and swift during floods. Canoes were usually used to reach the *pā-tuna*, although some had a walkway along the top of the fence. Most eel migrations take place over a few nights, so it was often necessary to empty the traps several times in darkness if the migration was a large one. Unfortunately, none of these very large *pā-tuna* exist today, although a number of smaller weirs are still used in a few places to ensure that traditional means of fishing are not lost. Fykenets have typically replaced the use of *hīnaki* as a means of capturing migrating eels.

Knowledge

As eels were such an important source of food, Māori developed an excellent understanding of their biology, and especially of how the catching of eels was affected by changing weather and seasons. Summer was the main season for feeding eels (yellow eels) to be captured, and large numbers were caught then and preserved as a source of food during winter. The *heke* months (March–June, depending on species and sex) were also particularly busy months, and Māori understood

Fig. 10.7 An eel weir (*pā-tuna*) used to catch migrating (silver) eels. Traps (*hinakī*) were set in the gaps along the wooden fence. Reproduced with permission of the Sir George Grey Special Collection, Auckland Libraries, AWNS–19080702–12–4

Fig. 10.8 A modern painting at Wairewa (Lake Forsyth) showing Tunaroa molesting the wife of Maui, who is shown with his spear in the centre of the painting

how catches varied by moon phase and water flow. *Heke* eels were sometimes described as rolling downstream together in a ball, which could be heard as it approached because of the slapping of the water by the eels' tails. It was also known that saltwater attracted migrating eels, and this knowledge was used when digging the channels at lake outlets where seawater filtered through the gravel.

Occasionally, Māori transferred small eels to inland waterways or lakes with irregular connections to the sea, to create new populations for subsequent harvesting. This was sometimes in the vain hope that new self-sustaining populations would be created, but other Māori knew better and had a saying that "unless the eel can get to sea it will not be found" (McDowall 2011).

To many Māori, the eel was and should be a creature of mystery. Consequently, Māori are often opposed to scientific studies tracking migrating eels to their spawning grounds and collecting or searching for leptocephali at sea. In his recent book on eels, James Prosek captured something of this when he wrote (Prosek 2010) "At the end of my trip to New Zealand … I felt as though I'd been immersed in a place where not knowing is something quite different from ignorance—where the unknown is tangible and sacred, whether it be the force of a water guardian, or the spawning place of the longfin eel."

Legends and Proverbs

According to one version of a legend, the wife of Maui, a demi-god and key figure in the myths and legends of Māori, was molested by Tunaroa, the ancestral eel (Fig. 10.8). Maui trapped and killed Tunaroa and cut him into four pieces, which were the origins of freshwater eels, conger eels, hagfish and lampreys, and frostfish. Similar stories are found throughout other Polynesian islands, and have parallels with the Biblical story of Adam and Eve where the snake beguiled Eve. A common addition to the story of Maui is that the severed head of Tunaroa gave rise to the first coconut tree (see the French chapter in this book for more information on that legend). Other stories in New Zealand speak of the ancestors of eels descending from the heavens during a severe drought, as a form of rain, and finding refuge in wetlands on earth. Although the stories differ according to tribe and across Polynesia, they have the common elements of eels being supernatural in origin and a gift (*taonga*) to people.

Māori had no written language, so history was passed down by word of mouth, and eels often feature in stories and legends. Many *tapu* (sacred) rituals and superstitions surrounding traditional Māori fishing relate to eels (McDowall 2011).Very large eels were often considered to be guardians of waterways (and sometimes referred to as *taniwha*) and

Fig. 10.9 An old carved eel that was found when a village covered by the eruption of Mount Tarwera in 1886 was excavated

were honoured—there are several accounts of these huge eels being able to bark like a dog or cry like a baby. Such large eels were often feared, and one type, *tuna tuoro*, was particularly feared because it had a reputation for seizing men and children wading across swamps; it was also claimed that it would pursue its victims across dry ground, and could only be deterred by setting fire to the grass and fern because the ash would then cling to its body and halt its progress. Some *taniwha* were reputed to be so large they had to swim out to sea to be able to turn around. Other eels were said to be extremely slimy and were regarded as special, because they led the migration (*heke*), and other migrating eels followed the slime trail. Although eels were primarily a source of food for Māori, they were highly respected and were occasionally adopted as pets and fed, and hence spared capture.

Eels also feature in many Māori proverbs, for example, *E moe an ate mata hi tuna, ara an ate mata hi taua* ("the eel-fishers' eyes sleep but those of the warrior are awake"), a reference to that fact that whereas an eel fisher can afford to relax, a warrior cannot. Some proverbs encapsulated aspects of eel biology, such as *He ua ki to po, he paewai ki te ao*, which means that a rainy night is good for eel fishing.

Eels (*tuna*) often feature in Māori place names. *Kai*, for instance, is the Māori word for food, so *Kaituna* indicates a place where eels were available as food; *wai* is water, so *Waituna* was a place where eels were found. Likewise, *roto* means lake and *awa* is river, so the common name of *Rototuna* for a lake indicates a lake with plentiful eels and *Awatuna* a river with lots of eels. Eels symbolized strength, but also food and nourishment, and were often included in carvings and paintings that adorn meeting houses and gateways. Indeed, a carved eel was found when a village buried by a volcano in 1886 was excavated many years later (Fig. 10.9).

Māori and Eel Management

Access to eel fishing areas was of great importance to Māori, and many settlements were close to such areas. Location of *pa* (villages) on the banks of rivers or the edges of lakes also meant that were the *pa* besieged by a hostile tribe, the inhabitants usually had access to eels for food. Wars were fought over the right to fish disputed waterways, and one of the most common Māori grievances from land purchase or confiscation by European settlers was the loss of fishing areas and weirs. Ownership of traditional fishing areas was often indicated by carved posts, and the poaching of eels was a serious offence sometimes punished by death.

As the most intensive eel harvest was of migrating fish, overfishing was probably not much of an issue historically. However, Māori sometimes used a concept of *rāhui*, which meant that an area was rested for a period to allow replenishment of stocks. If someone drowned in a river, the area was usually declared *tapu*, which meant no activity, including fishing, could take place until it was declared again suitable by a priest.

In recognition of the importance of eels to Māori, several lakes have been set aside as exclusively Māori fisheries (or at least, commercial fishing there is prohibited). Some of these agreements go back more than 100 years, although the past decade has seen more lakes, and the sections of some rivers, returned to Māori. As part of a redress process for past land confiscation, Māori have been given title to the beds of some large rivers and lakes, which means that they can be more effectively involved in managing those areas. There is also legislation that can declare particular reaches of rivers as non-commercial fishing areas, in recognition of their high customary importance to Māori. Under the founding partnership treaty between Māori and European settlers, the Treaty of Waitangi of 1840, Māori were granted continued access to fish resources, including eels. Further, to allow Māori access to the large quantities of eels required for special occasions such as family or tribal gatherings (*hui*) and funerals (*tangi*), special authorization can be given by nominated elders or guardians (*tangata kaitiaki*).

Commercial fisheries in New Zealand are managed under a quota management scheme (QMS) whereby a quota owner has continued rights to catch a certain quantity of fish. Regional harvest levels can be adjusted annually should there be indications of overfishing. Many Māori are opposed to the concept of continued (perpetual) ownership, because they regard resources as common rather than private property. Freshwater eels entered the QMS in 2000 (South Island) and 2004 (North Island), and as part of the settlement process, Māori were given 20 % of the overall commercial quota, plus a quantity (equivalent to 20 % of the commercial quota) available for customary harvest; ownership of a large processing plant for eels was also included in this settlement.

Since then, much of the North Island quota has been purchased by Māori companies, making Māori the largest owners of quota in the North Island. Now, because of concerns about the wellbeing of the longfin eel stock, some Māori choose not to fish for this species.

European History Regarding Eels

Relative to Māori, Europeans have had a very different understanding of the importance of eels. Usually, eels were regarded by Europeans settlers as pest species that were to be avoided. Early exploration and trade resulted in the destruction of many Māori eel weirs so that boats could navigate rivers more freely. Also, large areas of lowland wetland were progressively drained, and it is estimated that <10 % of all wetlands present when Europeans arrived in New Zealand a little over 200 years ago still exist (McDowall 1990). Such destruction of wetlands and river straightening has meant the loss of considerable habitat for eels, particularly shortfins. Being a species that favours flowing rivers and high country lakes, however, longfins have been more influenced by the building of weirs and dams, because historically no fish passes were provided past such constructions. Indeed, it was usually considered to be a good thing to prevent eels from gaining access upstream, because longfins preyed on the brown and rainbow trout introduced to New Zealand in the late 1800s to help British settlers acclimatize to life in New Zealand. Both trout species became established rapidly and spread throughout the country, and together they form the basis of New Zealand's legendary fishery for wild trout.

Eel Destruction Campaigns

Unfortunately, eels, especially the larger longfins, were known to prey on trout. This situation angered many of the anglers fishing for trout as well as the Acclimatization Societies (the managers of trout fisheries), who spent considerable time and money rearing juvenile trout in hatcheries and releasing them into the rivers around the country. Anglers were encouraged therefore to kill as many eels as possible, and in the 1930s, eel destruction campaigns were started in some regions; baited traps were used to catch as many eels as possible. An industry to extract oil from eels started during World War II but, although eels were still plentiful and their oil a rich source of vitamins, it failed, probably because cheaper sources of fish oil became available from overseas (McDowall 1994).

During the 1960s, although the intensive eel destruction campaigns had finished, many regions still had a bounty on eels, to encourage youth especially to kill them. Research was carried out then on the effect eels had on a trout population, but the results were a surprise to many trout anglers: when the eels were removed from a stream, the trout population increased almost ten times, although the average size dropped to a size too small to be of angling interest. The result therefore showed that the presence of eels, especially longfins, was useful in stopping trout from overpopulating an area. In fact, eels and trout now coexist in many of New Zealand's most highly regarded fisheries for wild trout, demonstrating that high quality trout fisheries can be maintained in the presence of eels. This knowledge effectively stopped the bounty system on eels.

Development of a Commercial Fishery

The commercial eel fishery in New Zealand started in the mid 1960s, mainly by immigrants from the Netherlands who were used to catching and eating eels in their homeland and saw an immediate export opportunity. The industry developed rapidly, and reached its highest level in 1972. Development of the fishery can be divided into stages (Jellyman 2007), with an initial exploitation stage (1965–1980), a consolidation stage (1980–2000) and a rationalization stage (2000 on). In the first stage, the industry expanded rapidly; there were many processing companies, almost anyone could get a permit to catch eels, and the product was high in volume but of low quality. During the consolidation stage, moves were made to slow the rate of growth of the industry by not issuing new catch permits and by developing regional management plans. In the rationalization stage, eels were introduced into the QMS, and many fishers left the industry. With less competition among fishers, eel fishing can be carried out now in a more sustainable way. All parts of the country can be fished except for National Parks and various reserves, and farmers rarely deny access to fishers. Together, such reserve areas, along with other areas that are seldom fished, mean that some 50 % of waterways and lakes where longfin eels are found are not fished commercially. There are both minimum (220 g) and maximum (4 kg) size limits on eels caught commercially, with the latter limit designed to protect a high proportion of female longfin silver eels. However, this upper size limit alone does not provide sufficient protection for large eels, because their slow growth rates mean that they are typically vulnerable to capture for more than 20 years before they attain a weight of 4 kg (Hoyle and Jellyman 2002). Modern fishing using fykenets is very effective, and the larger more-mobile eels are caught first. This leads to a reduction in the average size of eels in rivers (Jellyman 2009), something of concern to Māori because large eels are the traditional harvest and they have become far less available in many areas. There are also concerns about the ecological impact of removing the largest predator and scavenger from waterways.

Management of Eel Stocks

Management of New Zealand's eels is based on managing the harvest through the QMS, not allowing any harvest of glass eels or silver eels, having substantial reserve areas where no catching is allowed, and transferring small eels (elvers) from below hydroelectric dams to upstream waters. Therefore, whereas older hydroelectric dams had no facilities for the safe passage of eels, all large dams where elvers congregate now have a capture system so that elvers can be caught, counted and released at selected locations above the dam(s). At present, some five million small eels are transferred above the dams each year.

In the absence of any fisheries for glass eels, the records of elvers at dams are the only consistent measure of recruitment, and some regions are showing reduced recruitment of longfins. Hence, and because the species is unique to New Zealand, there are increasing calls to ban commercial fishing of longfin eels. In recognition of their reduced abundance too, some commercial fishers have decided voluntarily not to fish for them.

Because eels are nocturnally active, their access to daylight cover, where they can avoid direct sunlight, is crucial. Cover can be in the form of rocks and boulders, soft sediments, woody debris, undercut banks and aquatic plants. With the widespread river and stream channelling carried out throughout New Zealand to facilitate land drainage, however, considerable eel habitat has been lost. Although the uncontrolled growth of introduced willow trees (*Salix* spp.) can cause problems for river engineers, willow root mats provide excellent eel habitat. Willow removal and drain clearance remain major issues for maintaining eels stocks throughout the country, but it is unfortunate from a fisheries perspective that responsibility for aquatic habitats is held by agencies other than those that manage the fisheries.

Recommendations for regulation and research of the commercial fishery are made through annual meetings of an Eel Management Group consisting of fishery managers, Māori representatives, industry stakeholder groups, Regional Council representatives and fishery scientists.

Present Attitudes Towards Eels in New Zealand, and Research and Management

The feeding of "tame" eels has become popular at a number of places in New Zealand, because eels are easily attracted to an area at regular times and will eat a range of food. In some places, tourists pay to see and hand-feed the eels. There is growing recognition too of the ecological importance of eels, and children are fascinated by their complex life history, starting and ending in the oceans. There are broad concerns too that longfin eels are in decline, largely through habitat loss and restricted access, but also through overharvesting by commercial interests; longfins are currently listed among New Zealand's threatened fish species, though at a low level (Allibone et al. 2010). Whereas Māori have an inherent respect for eels, many Europeans dislike them, partly because they look like snakes (there are no snakes in New Zealand), but also because they grow so large and are attracted by splashing and the presence of blood in the water.

There are many stories too of eels attacking wounded fish, and of killing sheep stranded in ditches. Certainly large longfins do approach divers in lakes and rivers, but although most of these approaches are probably because the eels are inquisitive, some eels can be aggressive—one diver suffered a mild attack of the bends when he rapidly swam to the surface of a lake after being unnerved by the hostile way he was being approached by longfin eels estimated to be 36 kg in size. However, despite such occasionally bad publicity and stories about eels biting unwary swimmers, there is growing recognition of their importance and the unique place they hold in New Zealand's freshwater fish fauna. Several aquaria in New Zealand feature large longfins that are hand-fed (Fig. 10.10), and this is an important way that the public can be helped to understand and appreciate these fascinating fish. There is also a small but growing catch-and-release rod-and-line fishery for large longfin eels, mainly by European anglers, especially Scandinavians, fascinated by the opportunity to catch very large eels in scenic surroundings.

In terms of research, there has been considerable focus on eels over many years. It has included studies of the factors influencing recruitment, growth rate and abundance, comparisons of the diets and habitat preferences of all species, work on the ecological role of eels and also on the timing of downstream migrations. Research into the timing of downstream migrations is of particular importance, because maintaining adequate escapement of spawning eels to the sea is essential to the species' survival. Designing systems to divert migrating adult eels is particularly challenging, however, because eels tend to migrate during floods when classical diversion systems such as lights are less effective and netting usually impossible. Present-day research, therefore, is focused on tracking migrating (silver) eels equipped with acoustic tags, to determine whether there are times or locations when they are more susceptible to capture or diversion.

The future management of eels in New Zealand will clearly involve a closer working relationship between the fishery and land managers, Māori and commercial fishers. Until the turn of the century, Māori and commercial fishers hardly communicated, and many Māori still believe that there should be no commercial fishery for eels. Although there is interest in New Zealand in eel farming (and trial farms have been set up in the past), there is no farming

Fig. 10.10 A diver holding a longfin eel in a public aquarium (photo courtesy Southern Encounter Aquarium)

Fig. 10.11 A modern sculpture of longfin eels (photo and sculpture courtesy Bing Dawe)

Fig. 10.12 A schoolchildren's project on eels in their local stream. All the eels (57) in the net were released alive (photo courtesy Ian and Helen Perry)

currently and it is also illegal to export glass eels from New Zealand.

Apart from by Māori and other Pacific Island communities, eels are not widely consumed in New Zealand, although smoked eel is becoming a regular entrée in some restaurants. Eels are also starting to become more common in contemporary art and literature (Fig. 10.11), and occasionally feature in sculptures and paintings. Recently two episodes of the popular TV series "Monsterfish" and "River Monsters" featured New Zealand longfin eels. Also, a number of modern childrens' books feature eels and eel stories (e.g. Jones 2005), and in this way, younger generations of New Zealanders are starting to develop a greater understanding and respect for these amazing fish (Fig. 10.12).

References

Allibone R, David B, Hitchmough R, Jellyman D, Ling N, Ravenscroft P, Waters J (2010) Conservation status of New Zealand freshwater fish, 2009. N Z J Mar Freshw Res 44:271–287

Best E (1929) Fishing methods and devices of the Māori. Dominion Museum Bull 12:230

Downes TW (1918) Notes on eels and eel-weirs (tuna and pā-tuna). Trans Proc N Z Inst 50:296–316

Hoyle SD, Jellyman DJ (2002) Longfin eels need reserves: modelling the effects of commercial harvest on stocks of New Zealand eels. Mar Freshw Res 53:887–895

Jellyman DJ (1995) Longevity of longfinned eels *Anguilla dieffenbachii* in a New Zealand high country lake. Ecol Freshw Fish 4:106–112

Jellyman DJ (2007) Status of New Zealand fresh-water eels stocks and management initiatives. ICES J Mar Sci 64:1379–1386

Jellyman DJ (2009) Forty years on – the impact of commercial fishing on stocks of New Zealand freshwater eels (*Anguilla* spp.). In: Casselman JM, Cairns DK (eds) Eels at the edge, vol 58. American Fisheries Society Symposium, Bethesda, MA, pp 37–56

Jellyman DJ, Bowen M (2009). Modelling larval migration routes and spawning areas of Anguillid eels of New Zealand and Australia. In: Haro AJ, Smith KL, Rulifson RA, Moffitt CM, Klauda RJ, Dadswell MJ, Cunjak RA et al (eds) Challenges for diadromous fishes in a dynamic global environment, vol 69. American Fisheries Society Symposium, pp 255–274

Jellyman DJ, Tsukamoto K (2010) Vertical migrations may control maturation in migrating female *Anguilla dieffenbachii*. Mar Ecol Prog Ser 404:241–247

Jones J (2005) New Zealand freshwater eels. Reed Children's Books, Auckland, 40 pp

McDowall RM (1990) New Zealand freshwater fishes: a natural history and guide. Heinemann-Reed, Auckland, 553 pp

McDowall RM (1994) Gamekeepers for the nation. The story of New Zealand's acclimatisation societies, 1861–1990. Canterbury University Press, Christchurch

McDowall RM (2011) Ikawai. Freshwater fishes in Māori culture and economy. Canterbury University Press, Canterbury, 872 pp

Prosek J (2010) Eels. An exploration from New Zealand to the Sargasso, of the world's most amazing and mysterious fish. Harper-Collins, New York, 287 pp

Rowe DK, Chisnall BL, Dean TL, Richardson J (1999) Effects of land use on native fish communities in east coast streams of the North Island of New Zealand. N Z J Mar Freshw Res 33: 141–151

Schmidt J (1925) On the distribution of the fresh-water eels (*Anguilla*) throughout the world. 2. Indo-Pacific region. Kongelige Danske Videnskabernes Selskabs Skrifter 8:329–382

The American Eel: A Fish of Mystery and Sustenance for Humans

11

Michael J. Miller and John M. Casselman

The American eel, *Anguilla rostrata*, is, in many ways, a typical anguillid in body shape and size and life cycle (Fig. 11.1), and also in that people who encountered it and needed sustenance always found it to be excellent food. It lives in watersheds adjacent to the Atlantic Ocean from the Caribbean Sea and Gulf of Mexico to Atlantic Canada (Fig. 11.2) and it has been caught and eaten by humans in all parts of its range at one time or another. The American eel has been important in the lives of many people historically and still is today, but its importance has been largely unnoticed by many. Probably as much as for any anguillid eel in the world, however, anthropogenic activities and particularly dam-building have reduced the area in which the species once lived (Fig. 11.2).

Native peoples and early colonial settlers along the entire east coast of North America knew the American eel to be an important source of sustenance and protein. Eels were valuable as food on both sides of the Atlantic far back in human history (Casselman 2003; and see other chapters in this book), but little else about them was certain. Then, early in the twentieth century, both American and European (*Anguilla anguilla*) eels became famous among biologists and the general public as much for their mystery as for their amazing migrations. For centuries, if not for millennia, humans wondered where the eels went to reproduce. Reproductive-stage silver eels would migrate out of lakes and rivers into the sea in autumn, when many of them were caught for food, but where they went to spawn remained unknown. Then, as surely as the seasons change, there would be a rush of young glass eels returning to swarm up the rivers and streams after winter. These would then disappear into their watery world, rarely to be seen as they develop into silver eels. This was the life of eels as seen by humans, but what they did in the ocean was still unknown.

Then, in 1904, Johannes Schmidt and colleagues from Denmark collected the first leptocephalus (the technical term for an eel larva) of the European eel in the North Atlantic west of the Faroe Islands (Schmidt 1927). Catching that 77-mm long leptocephalus was significant, because there was a belief at the time that the European eel reproduced in the Mediterranean Sea, so its capture off the Faroes provided the first clue that these mysterious fish might spawn in the Atlantic. Its collection triggered an 18-year quest to find the birthplace of eels, a quest that extended to other species in other oceans (Schmidt 1935) and seemingly stimulated other scientists, including some studying the American eel, to search the oceans for eels. Schmidt collected leptocephali all over the North Atlantic, but the smallest leptocephali of both American and European eels were captured only in the southern Sargasso Sea; it was clear that both species must reproduce only there (Schmidt 1922).

This discovery that both European and American eels swam thousands of kilometres out into the ocean to reproduce seemingly brought these unglamorous, secretive fish to the attention of the world as never before. They were familiar to Native Americans in many parts of eastern North America because they provided more nourishment than many other fish (Casselman 2003). Others later used aboriginal methods of building weirs to catch downstream-migrating silver eels, and even today such weirs are still used in some places (Prosek 2010). Meanwhile, the story of the incredible migration of eels out into the ocean and the return of the larvae to where their parents had lived on two widely separated continents became a subject of huge interest to many scientists.

Other efforts were made to learn more about the spawning areas of American and European eels (McCleave 2003), and research into their biology in continental waters continued throughout the twentieth century. There was another change

M.J. Miller (✉)
Atmosphere and Ocean Research Institute, The University of Tokyo, Kashiwanoha, Kashiwa, Chiba 277-8564, Japan
e-mail: miller@aori.u-tokyo.ac.jp

J.M. Casselman
Department of Biology, Queen's University, 2406 Biosciences Complex, 116 Barrie Street, Kingston, ON, Canada K7L 3N6
e-mail: john.casselman@queensu.ca

K. Tsukamoto and M. Kuroki (eds.), *Eels and Humans*, Humanity and the Sea,
DOI 10.1007/978-4-431-54529-3_11, © Springer Japan 2014

in how people viewed the American eel when the numbers of glass eels or juveniles returning from the sea to freshwater suddenly declined, especially the numbers counted as they ascended eel ladders at dams on the St Lawrence River (Castonguay et al. 1994; Casselman et al. 1997a; Casselman 2003). These dramatic declines of the once large eel stock of the upper St Lawrence River and Lake Ontario (Casselman et al. 1997b) and also the declines of European (Dekker 1998) and Japanese (*Anguilla japonica*) eels (Tsukamoto et al. 2009) prompted the holding of a scientific symposium specifically on anguillid eels in Québec, Canada, in 2003 (Casselman and Cairns 2009); scientists at that meeting publicly expressed their concern strongly (Dekker et al. 2003). Many reasons were proposed for the declines, but identification of the main reason remained elusive.

Fig. 11.1 The American eel, *Anguilla rostrata*, at the yellow, natural silver and artificially matured silver eel stages (*top to bottom*). The larger eyes and darker body colouration are characteristics of silver eels migrating to their spawning area in the Sargasso Sea. Photograph courtesy A. J. Haro

American eels traditionally moved up the large rivers far into the North American continent, but they were losing access or safe passage to vast areas of their former range (Fig. 11.2) as a result of dam construction for hydroelectric power or flood control (Haro et al. 2000). This reduced the habitat available to them, and even if the young eels could still climb over small blockages or use eel ladders constructed to help them, downstream-migrating silver eels experienced heavy mortality at the dams when attempting to return to sea (Haro et al. 2000; McCleave 2001).

The decline of eels in North America and the northern hemisphere generally began to become an increasingly interconnected global issue for humans because the Japanese eel was traditionally very important as food in East Asia, with demand greater than the supply from the wild. Intensively cultured, with wild-caught glass eels used as seedlings for producing market-sized eels (Tsukamoto et al. 2009), the

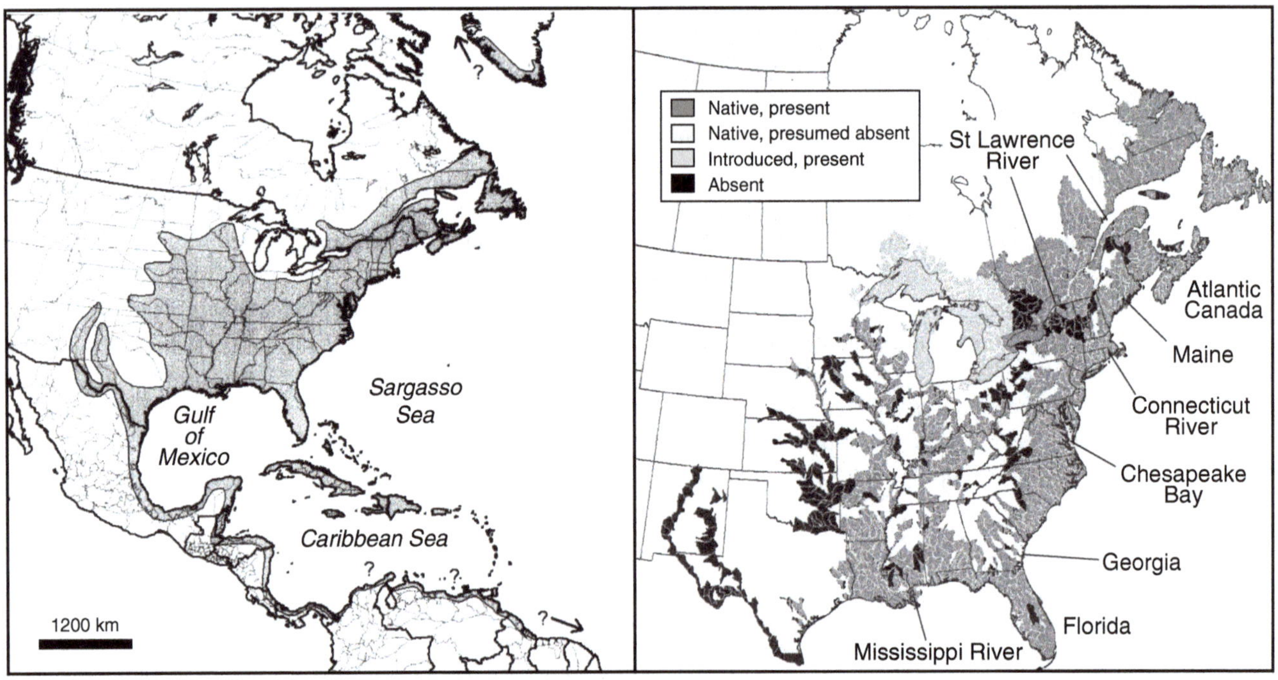

Fig. 11.2 Maps showing the historical distribution range of American eels in North America (*left*), and the areas where they still remain through natural recruitment or stocking and where they are no longer present because of blockages to upstream migration by dams (*right*). Adapted from USFWS (2007)

decline in recruitment of Japanese eels forced glass-eel traders to look to North America and Europe for another source of their seedlings. This resulted in a "goldrush" for glass eels in places such as the US state of Maine that still allowed glass eels to be fished (Facey and Van Den Avyle 1987; Prosek 2010). The price paid for glass eels soon rocketed, and guns were drawn as fighting ensued. In spring 2012, the price paid by distributors for glass eels reached an all-time high of US$2,600 per pound, and during the 2013 season for glass eels, there were daily newspaper articles about rampant illegal fishing. Indeed, in a television documentary aired that year, individual fishers in Maine were stated to be making up to $104,000 in a single night (Public Broadcasting System, PBS, documentary, *The Mystery of Eels*). Apparently too, on the opening night of the 2012 season, which had been on a Thursday, a local eel dealer paid >$250,000 in cash and IOUs to local fishers for their catches of glass eels, and all the banks in coastal Maine ran out of cash and had to wait until the following Monday to receive more from Boston. The problem arose simply because glass eels from all over the world were being sought to support the extensive aquaculture industries in China, Japan, Taiwan and Korea (see other chapters in this book).

Unlike the Japanese eel, the American eel has not traditionally been very important as food for most Americans and Canadians, except for Aboriginals, so many of the silver eels historically caught in places such as Maine and the St Lawrence River system were shipped to Europe, where the tradition of consuming eels was stronger. Eels can be packed in high density and shipped live or frozen. However, that market declined around 1982 because of economic factors (Facey and Van Den Avyle 1987).

Scientists have long realized the ecological importance of the American eel and extensive research has been conducted on the species. Realization that its population was declining resulted in more dedicated research being carried out, new regulations being established to protect it, and systems to monitor its population being expanded (ASMFC 2012). Much of the research focused on determining how to facilitate glass eels ascending and silver eels descending past dams safely (Richkus and Dixon 2003). The decline in the population even triggered efforts to have the species declared as endangered in the US (USFWS 2007; Prosek 2010).

The American eel has played an important part in the lives of the people who share the rivers, lakes and streams with the species. It has inspired scientists to devote their whole lives to studying the species and to spend years at sea on research ships. Before being stopped by anthropogenically constructed barriers, the species travelled as far inland in the North American continent as any other anguillid eel on another continent. Despite being distributed from the Caribbean to Atlantic Canada, the American eel is known to be a single randomly mixing population (Wirth and Bernatchez 2003), so this chapter now looks briefly at some aspects of the relationships between American eels and the people who lived alongside them.

Eels as Food in North America

Casselman (2003) overviewed evidence of the importance of the American eel to Aboriginals in the St Lawrence River region of northeastern North America, and the evidence was revisited by MacGregor et al. (2012). Various tribes clearly used eels as a food resource. The St Lawrence River system extends far inland and includes many tributaries and Lake Ontario, and Junker-Andersen (1988) provided a unique ethnic and archaeological perspective on the lives of Iroquoians and their fisheries for eels there. Smoked eel was light in weight but nutritious, making it an important food for travellers (Casselman 2003). Indeed, it is likely that eel meat is more nutritious than that of most other fish, so it provided a valuable source of protein and helped people survive the long, cold winters (Casselman 2003). Large numbers of eels could be speared at night, using fire as a light source, in tributaries of Lake Ontario (Casselman 2003). Runs of migrating silver eels started in late summer or autumn, and these eels could be caught in large numbers in weirs constructed across streams or within large rivers, providing a valuable resource just before the onset of winter. Smoked, stored eels were used to prevent famine if other food was not available.

The extent to which American eels were used by other native peoples farther south, in warmer areas of the species' range, appears to be less well documented. An early woodcut from 1590 by Theodore de Bry shows the use of spears and weirs to catch fish, including eels, in Virginia (Fig. 11.3). It would be surprising, however, if eels were not being utilized as at least a seasonal food source by all the peoples who lived within their range. Indeed, in spring 1621, soon after the Pilgrims landed at Plymouth Rock in Massachusetts, there was a desperate need of food, and the first thing a local chief who decided to help them showed them was to catch and eat eels, probably saving them from starvation (Prosek 2010). For this reason, some have suggested that eels be made the symbol of the American Thanksgiving dinner rather than turkey!

Eels are still smoked and sold in some parts of the US, and eel is consumed in small quantities in Atlantic Canada (Fig. 11.4; Prosek 2010), but the eels of the east coast of North America are still not really a familiar or common food of modern Americans and Canadians. Perhaps many, especially those living outside Atlantic Canada and the eastern US, have never eaten eel, and some of those living along the coast may be more inclined to think of small eels less as food and more as bait for catching fish such as striped bass.

Fig. 11.3 Orr's (1917) reproduction of an engraving from 1590 by Theodore de Bry in Virginia showing the use of spears and weirs to catch fish including eels. Reproduced with permission from the Archives of Ontario, Toronto, Ontario

Fig. 11.4 Sliced and seasoned smoked eel packaged for sale (*left*) and a sign advertising the sale of eel and other fish in Québec, Canada. The listing of "anguille" for eel in French is referring to American eel

The small eels there are caught in eel pots in rivers or estuaries and sold to fishers as bait. Until recently too, many of the migrating silver eels captured in weirs in rivers and streams in out-of-the-way locations in Maine (Fig. 11.5b) and to the south were gathered by distributors who air-freighted them to markets in Europe and, more recently, Asia. Prosek (2010) provides examples of these weirs and the men who fish them, and the way one fisher catches silver eels in a weir (Fig. 11.5a) on the east branch of the Delaware River near Hancock, New York State, in the Catskill Mountains was also featured in the documentary on eels aired by PBS in their Nature series, *The Mystery of Eels*: throughout summer, the long stone walls of the weir are repaired while the water of the river is low, to prepare for the autumn downstream migration of silver eels. Most of the run is typically over a period of just a few nights during a new moon, when a storm in the catchment causes the river level to rise, and >1,000 eels can be caught in the weir in a single night, to be

Fig. 11.5 Examples of eel weirs used to collect silver eels of the American eel in recent years: (**a**) the eel weir built by a fisher on the East Branch of the Delaware River in the Catskill Mountains of New York that was featured in the recent book and documentary about eels (Prosek 2010; PBS documentary: *The Mystery of Eels*; photo courtesy J. Prosek), (**b**) an eel weir on the Seboeis River in Maine (photograph courtesy A. J. Haro), and (**c**) an eel weir on the St Lawrence River upstream of Québec City, Canada. The size of the weir can be seen relative to two pick-up trucks parked nearby. The eels are removed from the weir during low tide

smoked and sold locally. To the north too, at the mouth of the St Lawrence, there was a long tradition of fishing huge tidal weirs that extended out from the edge of the river (Fig. 11.5c; Verreault et al. 2003), although typically those fishers were farmers trapping the large eels in autumn as an extra source of income.

More recently, young North Americans have become familiar with eels because of the greatly increased popularity of Japanese sushi restaurants. However, despite American eels being found along the whole east coast of North America, those consuming sushi with eels in North America are more likely to be eating American or European eels caught as glass eels in Maine or France, then shipped to China to be raised before being returned to North America for serving on sushi.

Spawning Areas

It is difficult to know how the general public viewed the American eel during the first decades of the twentieth century when Johannes Schmidt's research expeditions discovered the spawning area of both American and European eels in the southern Sargasso Sea. Schmidt did perhaps envision the importance of knowing where eels were reproducing though, because he used many ships to collect the samples he needed to determine where European eels were spawning, including research vessels, ships of opportunity, particularly Danish commercial ones, and naval vessels (Boëtius and Harding 1985; McCleave 2003). Johannes Hjort, on board the steamer "Michael Sars" also caught small eel larvae in the

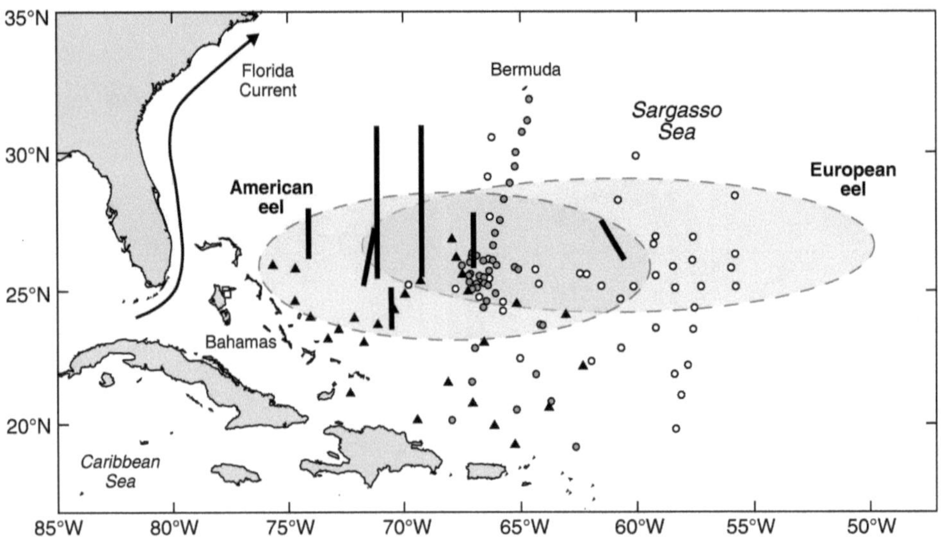

Fig. 11.6 Map of the Sargasso Sea spawning area of American and European eels, showing the locations where tows of Isaacs–Kidd midwater trawls were made by American and German research teams between 1979 and 1989. Circles depict the German stations (Schoth and Tesch 1982, *white circles*; Tesch and Wegner 1990, *grey circles*), triangles the stations of Kleckner et al. (1983), and lines the transects of Kleckner and McCleave (1988), except for the westernmost transect that was made during the 1989 expedition to locate and catch adult eels. The shaded ovals show where the smallest leptocephali of each species were collected (<6 mm long) during all cruises by Danish, German and American research teams up to 2007 (adapted from unpublished material of the first author and others)

central Atlantic, suggesting that the area where eels spawned might be far out in the Atlantic (Hjort 1910).

Much is written about Schmidt's life and work in the Danish chapter in this book, but to supplement that and to focus more on the American eel, further detail is given here. Schmidt initially used the schooner "Margrethe" to travel from Europe to the West Indies in 1913, and encountered both American and European eel leptocephali in some of the same tows of his sampling net (Boëtius and Harding 1985). However, later that year the "Margrethe" was wrecked on Anegada Island, British Virgin Islands, and with World War I intervening, no more collections of eel larvae were made until 1920 (Boëtius and Harding 1985). Schmidt then returned to the Sargasso Sea in 1920 in the motor schooner "Dana I," to the area where the smallest leptocephali had been collected before the war, and he caught many more between February and August (Schmidt 1922; Boëtius and Harding 1985). It was the data collected in those surveys that led to Schmidt's description of the spawning area of American and European eels (Schmidt 1922), but further collections were made by the "Dana I" from February to May 1921 and by the RS "Dana II" from April to June 1922, and these were consistent with the published conclusions about the spawning areas of the two species (Boëtius and Harding 1985; Kleckner and McCleave 1985; McCleave 2003). The implications of Schmidt's discovery for scientists and the general public were remarkable: both species were migrating thousands of kilometres through almost featureless ocean to the same small area of the North Atlantic to spawn, but there was no understanding of how they could possibly achieve this. Equally remarkable from the standpoint of basic logic was how the small leptocephalus larvae that originated from almost the same area of the Sargasso Sea could return to their respective habitats in either North America or Europe without becoming totally mixed in each area.

Although Schmidt's work publicized both species of eel for their migrations each way, it was clear to scientists in both North America and Europe that there was much more to learn about the spawning areas of the two. All Schmidt's data plus other data on the American eel were then assembled and eventually published to provide the first perspective of what had been collected by Schmidt and others (Kleckner and McCleave 1985). The first American research expedition to the spawning area was led by James D. McCleave of the University of Maine in February 1981. That survey detected a temperature front and collected leptocephali along its southern edge (Kleckner et al. 1983; Fig. 11.6). Independently 2 years earlier, a German team led by Friedrich-Wilhelm Tesch had collected leptocephali over a much wider area. Tesch was well-known at the time for publishing his comprehensive work "The Eel" in 1977, and it was updated and reprinted in 2003 (Tesch 2003). The German expedition had hoped to collect eel larvae with plankton nets and adult spawning eels with a midwater trawl. They documented how American and European eel larvae were spread across the longitudes of the spawning area (Schoth and Tesch 1982), but no adult eels of either species were caught. The German

Fig. 11.7 Cover photographs of the journals *Science News* (cover photograph made by the first author, reprinted with the permission of *Science News*) and *Nature* (cover photograph by N. Mochioka, permission granted by Nature Publishing Group) that included articles about the American expedition to the Sargasso Sea spawning area to collect spawning adult American eels in 1989 and the discovery of the spawning area of the Japanese eel in 1991

team made a second expedition in 1981 to determine the latitudinal distribution of larvae (Tesch and Wegner 1990; Fig. 11.6), and the American team continued its research with cruises in 1983, 1984 (for large larvae), 1985 and 1989 (Fig. 11.6). The cruises in February and April 1983 and in March 1985 consisted of transects crossing the temperature fronts thought to influence the spawning locations of eels (Kleckner and McCleave 1988), and the catch data revealed that a northern front in the spawning area formed the northern limit of spawning (Kleckner and McCleave 1988). Finally, the collections of larvae by Schmidt and the German and American survey teams were combined to evaluate the overlap in the spawning areas of the two eel species based on the small leptocephali collected (McCleave et al. 1987; Fig. 11.6).

The most recent American expedition to the spawning area was made in 1989. Its goal was to use advanced fish-finder-type hydroacoustic systems to search for spawning aggregations of eels near the fronts and to attempt to capture them by trawl or in traps baited with artificially matured American eels. Many fish-like targets were seen on the echosounders that could have been eels, but not one eel was caught. The story of that cruise was published in Science News Magazine (Pennisi 1991), shortly after another cruise targeting eel larvae in the western North Pacific had succeeded in locating the spawning area of the Japanese eel (Tsukamoto 1992), a goal of Japanese scientists for decades. The story contrasted the outcomes of the two expeditions, one with massive effort to find and catch spawning adults that ended in disappointment, the other that succeeded well beyond expectations and which was symbolized by a cover photograph of a rainbow taken during the cruise in the Pacific (Fig. 11.7). Although the American survey was unsuccessful at catching any adult eels, another transect was made to collect larvae, and the data from that transect were combined with the catch data from other surveys to investigate the overall assemblages of leptocephali in the Sargasso Sea (Miller and McCleave 1994).

Despite these successes, however, it was another 20 years before scientists finally collected adult anguillids and their recently spawned eggs in the ocean; these were of the Japanese eel in the Pacific (Chow et al. 2009; Tsukamoto et al. 2011).

In many ways, the quest by research teams from Denmark, Germany and the US to understand the reproductive ecology of the Atlantic eels, and the many research expeditions by Japanese scientists in the Pacific (Shinoda et al. 2011), represent a remarkable example of scientific resolve, perhaps to the point of obsession, to learn all there is to know about these mysterious fish. One part of the quest was that of Katsumi Tsukamoto of the Ocean Research Institute of the University of Tokyo, one of the editors of this book, who made many surveys that ultimately led to the collection of anguillid eel eggs for the first time on 22 May 2009 (Tsukamoto et al. 2011), 105 years to the day after Schmidt's collection of the first anguillid leptocephalus in the Atlantic Ocean on 22 May 1904. The research is continuing to this day, with a recent Danish survey to the Atlantic spawning area in 2007 (Munk et al. 2010) followed by a German survey in 2011; both teams are also planning surveys again in 2014. Surveys in the Pacific have now switched to attempting to observe spawning eels directly in the wild during New Moon periods, using submersibles and other underwater camera systems (Tsukamoto et al. 2013).

The American Eel in Rivers, Lakes and Estuaries

The life history of the American eel is similar to that of other anguillids, and many of the fascinating discoveries about eels have been made by those studying *A. rostrata*. It has the same life history stages of egg, pre-leptocephalus, leptocephalus, glass eel, yellow eel and silver eel as other species of eel, and its adult size is similar to that of most other species, though smaller than some of the tropical ones. Although its spawning migration is not as long as that of the European eel, its migration is still very long, much longer than that of some tropical eels. The American eel is also apparently good at keeping time, having what is known as a biological clock. The generally transparent glass eels enter estuaries (Haro and Krueger 1988), but most then must move upstream, which in large rivers can be difficult because of the outward flow. Studies on glass eels entering the Penobscot River estuary in Maine provide some of the best information about the mechanism of selective tidal stream transport used by eels and other fish. In large estuaries with strong tidal flow, such as the lower Penobscot, glass eels swim upstream during flood tides, using the flow to help them up the river (McCleave and Kleckner 1982). They also seemingly rest on the riverbed during ebb tides, and the timing of their movement off the riverbed is regulated by a biological clock related to the tidal/lunar cycle (Wippelhauser and McCleave 1988).

The young eels entering and moving up rivers are rarely seen, though, except by those who inspect small dams during spring and see the eel-like creatures slithering up wet areas and over the obstruction. Once the young eels find a place to settle and start feeding, their yellow eel juvenile growth stage begins, and then they are encountered only by those who catch them with a hook baited with a worm or minnow and leave them to lie on the bottom of a stream (Prosek 2010). An angler may at first mistake an eel for a large bass or another fish, because eels are able to swim strongly backwards, pulling the line very hard.

American eels emerge from their daylight hiding place at night to feed on aquatic insects, crustaceans, small molluscs, frogs and small fish (Facey and Van Den Avyle 1987), just like other eels in similar habitats. Research on the feeding behaviour of American eels has revealed that when they encounter food items too large to swallow, they grasp the food in their jaws and spin their bodies round and round rapidly to tear off a piece to swallow (Helfman and Clark 1986). This has been referred to as "rotational feeding," and smaller eels have been recorded spinning their bodies at rates of up to 14 turns per second! The time American eels spend feeding and growing as yellow eels before they start their migration back to sea can vary widely, from several years to decades, based on the habitat in which they live, the density of other eels present and the latitude (Oliveira 1999; Oliveira and McCleave 2000; Jessop 2010). In fact, the sex of each American eel and likely all other anguillids seems to be determined by the density of eels and the growth conditions (Krueger and Oliveira 1999). Male yellow eels mature younger than females, which also attain a larger size, and this is likely part of their reproductive strategy (Helfman et al. 1987). American yellow eels tend to find an area in which to live and feed without moving around much, i.e. a home range, and the use of electronic tags has confirmed that they do tend to stay in that range, be it a stream, pond or other habitat, including river mouths and estuaries. One study showed that yellow eels used selective tidal stream transport to move small distances up and down the lower Penobscot River to feed before returning to their daytime hiding place (Parker 1995).

As mentioned elsewhere in this book, it has been possible to determine from the strontium/calcium content of the otoliths (ear stones) of American eels whether an eel has lived in totally freshwater or in brackish water in an estuary. Many American eels that entered freshwater as glass eels and lived there for some time then returned to the estuary to finish their yellow stage, whereas others remained in the estuary seemingly without ever having entered freshwater (Jessop et al. 2008).

Anyone who has ever visited an eel weir on a river or stream such as those shown in Fig. 11.5a, b after a rainy night will carry away a feeling of the wonder of migrating eels for the rest of their lives. A surprising variety of sizes, shapes and colours of eels from anywhere upstream of the weir can all end up in the same catch box at the same time, despite having developed in very different habitats upstream. Some will have come from just upstream of the weir and some from hundreds of kilometres away, perhaps in a small stream or lake.

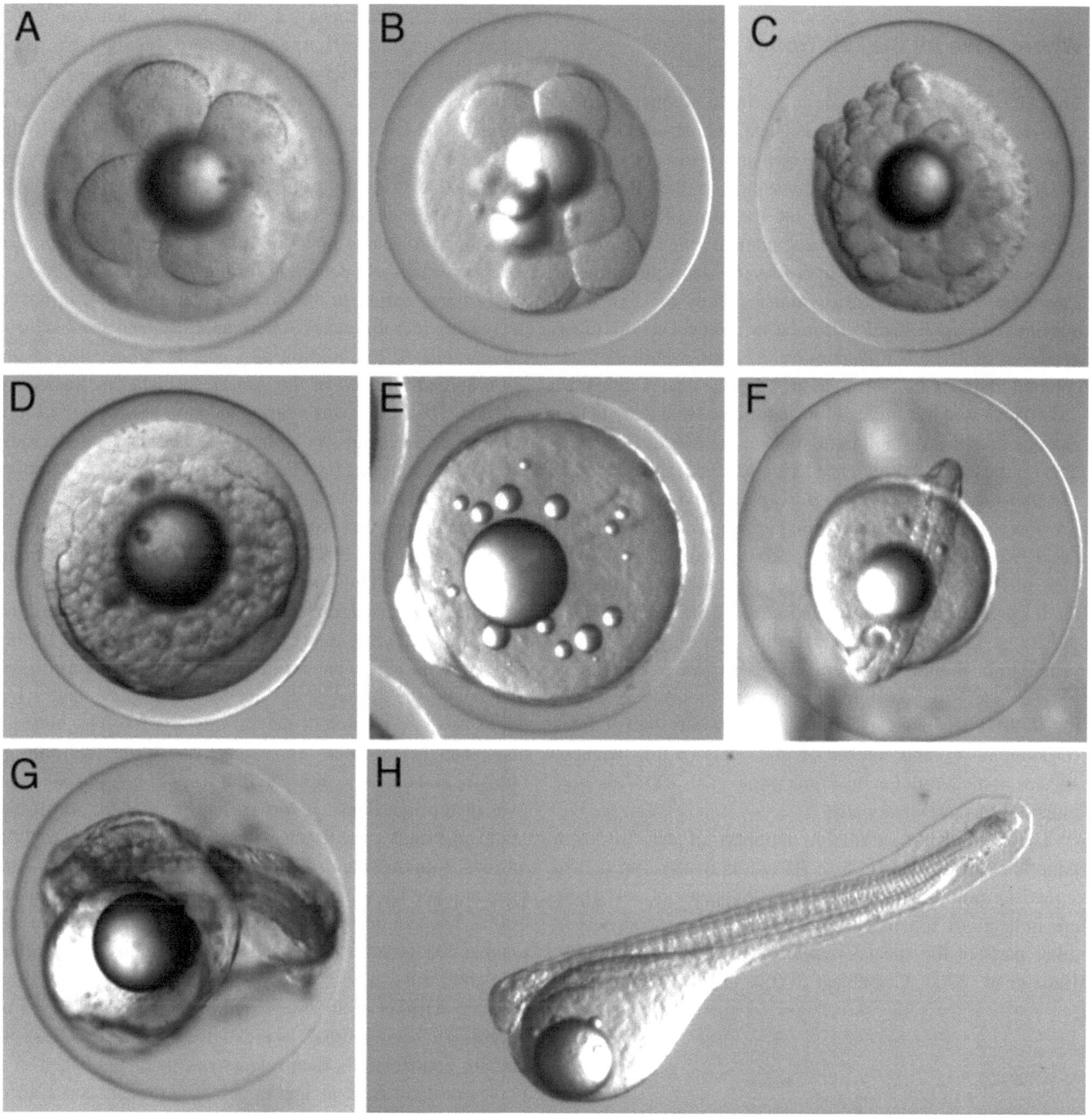

Fig. 11.8 American eel eggs (embryos) obtained from artificially matured silver eels caught in the Sebasticook River, Maine, and held in the laboratory in Massachusetts (reproduced with permission from Oliveira and Hable 2010. See title and publication detail in References. © Canadian Science Publishing or its licensors)

Downstream migrations generally take place during autumn as temperatures fall and are usually complete before winter. American eels and other temperate anguillids tend to migrate downstream on rainy nights or during the dark phase of the lunar cycle (Haro 2003). Any master of an eel weir will report that few eels are caught on a clear night with a full moon even in the heart of the autumn migration, but if there is a big storm during the night of a full moon, introducing a lot of precipitation into rivers and streams in an area, more eels will be caught than during a clear night under a new moon.

Some have recently unlocked parts of the secretive life history of the American eel by inducing silver eels to spawn eggs in the laboratory (Oliveira and Hable 2010). This was accomplished by injecting the eels with hormones to stimulate their gonads to develop even though they were not migrating through the ocean. Without such injections, eel gonads will not mature in captivity. Such techniques were pioneered in Japan (see the Japanese chapter) and have been effective in producing eggs that have hatched into American eel larvae (Fig. 11.8).

Where Have All the Eels Gone?

The American eel is fortunate in having a number of long-term data series to index its abundance and recruitment, particularly in the St Lawrence River and upstream in Lake Ontario. The indices provide information on recruitment and harvest data provide evidence of the size of the exiting spawning stock. An important recruitment index is based on counts at eel ladders on the hydroelectric dams along the St Lawrence. As small eels accumulate below the first dam, densities increase, likely stimulating many eels to move upstream over the fish ladders on the dam, where they are counted daily. Casselman et al. (1997a) and Casselman (2003) reviewed this information and showed that daily ladder passage of upwardly migrating eels at the Moses–Saunders dam on the upper St Lawrence River declined from a peak in 1982/83 to a very low value during the 1990s (Fig. 11.9). The few eels that did ascend were larger and older, but by the late 1990s, the numbers had dropped to virtually zero. Since then, however, a few eels have passed up the ladders, but still far fewer than in the 1980s and earlier (MacGregor et al. 2012; Fig. 11.9). The dramatic decline over the past 20 or so years is not limited to American eels, however. Catches of glass eels at a long-standing recruitment-monitoring site in the Netherlands have also dropped to virtually zero at times (Dekker 1998), and catches of glass eels of the Japanese species in the North Pacific (Tsukamoto et al. 2009) are now much lower than historically.

These dramatic declines in numbers of eels in North America and elsewhere have served as a wake-up call to scientists monitoring the status of all species of eel. There are indications too that there are fewer American eels in other parts of the species' range (Casselman et al. 1997b; Haro et al. 2000; Casselman 2003), though without such long-term datasets, there is no other equally clear image of a species disappearing from a whole part of a continent. If no eels climb the ladders of the massive dams on the St Lawrence, there will eventually be no eels in the watershed of rivers and lakes upstream of those dams, including in the huge Lake Ontario. Another concern is that the St Lawrence watershed historically supplied large numbers of the big female silver eels migrating back to their spawning area, so with those so scarce now, the total spawning population of the species is likely declining further.

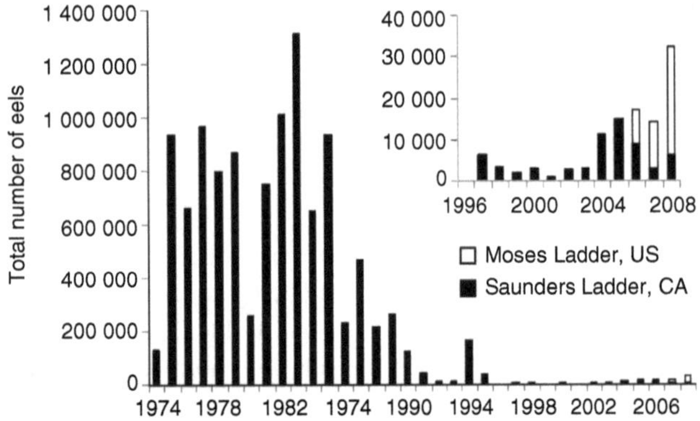

Fig. 11.9 Trends in the catches of American eels at the two Moses–Saunders Dam fish ladders, showing the sudden declines in numbers of eels going up and over the dams, and a small increase in numbers in recent years (inset). Modified after MacGregor et al. (2012)

As stated earlier in this chapter, alarm at the continually decreasing recruitment of American, European and Japanese eels stimulated the holding of a special symposium at the 2003 annual meeting of the American Fisheries Society in Québec. That venue was a fitting location for the symposium because at the "narrows" where Québec City now stands, First Nations groups (see below) historically had assembled annually in autumn to negotiate and to catch eels in the St Lawrence River that were present in "an almost unlimited supply" (du Creux 1664), and which "cost nothing beyond the catching" (Thwaites, 1896–1901). That symposium brought together experts on anguillid eels from all around the world, including many of the authors of chapters of this book, to consider what needed to be done to identify and understand the cause of the declines and to evaluate what could be done to protect the various species. The meeting also brought together representatives of the hydropower industry and others working to find ways of helping eels pass over or through dams. It was perhaps one of the most dramatic meetings on eels ever held in North America, with various opinions being expressed on how to approach the problem (see the Panel Discussion in Casselman and Cairns 2009). The synthesized discussions led to the signing of the Québec Declaration of Concern for the status of anguillid eels around the world, which was published soon after the meeting (Dekker et al. 2003), and the story was featured in several articles in journals as prestigious as *Nature*. The symposium contributions themselves were later published in a book entitled "Eels at the Edge" (Casselman and Cairns 2009; Fig. 11.10). However, there was an unexpected end to the symposium for which the focus had been the struggle of a group of fish species that had existed unchanged for millions of years, that were now simply trying to swim upstream to their ancestral feeding grounds and then to return downstream back to their birthplace. Humankind needs a steady, affordable source of electricity such as from dams, and the day after the symposium ended, the great northeastern electricity blackout struck, cutting the power to millions of people in the northern and eastern US and large regions of Canada. Flights to and from Québec were cancelled and schedules had to be changed. Remarkably, Québec City, the site of the symposium, did not lose power, but it is certain that participants did not fail to see again how society depends

Fig. 11.10 Front and back covers of the Casselman and Cairns (2009) book "Eels at the Edge. Science, Status, and Conservation Concerns" that resulted from the Québec eel symposium in August 2003, when scientists released the "Québec Declaration of Concern" (Dekker et al. 2003)

so much on electricity. However, their resolve to protect and better understand the American eel and other anguillid eels was strengthened by their experience.

One way to address the issue has been to direct research at the factors influencing whether or not eels can safely pass up over or back down through hydroelectric dams (Richkus and Dixon 2003). Various ladder designs had been tried, but eels would pass over some and not others. It was necessary to know, therefore, what combination of location relative to water flow below the dam and what configuration of water flow and gradient inside the ladder would attract the eels and encourage them to move up and over the dam. Similarly, if down-migrating silver eels could be attracted somehow to a particular part of the dam, bypass channels could be provided that would allow them to pass over the dam rather than through the turbines. Scientists and engineers explored a variety of ideas ranging from screens in front of the intakes to producing noise or light to scare the eels away from the intakes. Scientists tagged and tracked eels as they approached the dams to learn about their behaviour and how they reacted to various aspects of the dam or diversions to spillways and passages (Haro and Castro-Santos 2000). However, the eels tended to follow the main water flow into the turbines and were not easily guided by mechanical or behavioural means. Like many species, the huge variation in behaviour of individual eels makes them a difficult creature to control by any method, so the problem of helping them pass over or through dams remains a challenge.

Dams are not the only problem facing the American eel. Loss of habitat has followed the filling-in of wetlands, and overall societal development, pollution and agricultural runoff have reduced the quality of the habitat used by American eels across much of their present range. The decline in recruitment seemed to start after about 1975, corresponding to a major regime shift in the North Atlantic Oscillation (NAO), an atmospheric index related to climate patterns in the region (Hurrell 2005). This led to the hypothesis that some parameter in the ocean might have changed, lowering recruitment, given that immediate spawning success and the survival of larvae likely influence annual recruitment

(Knights 2003; Friedland et al. 2007; Miller et al. 2009). The relationship between the atmosphere and the ocean can affect the productivity of plankton, which could then influence the survival of leptocephali as they are carried in currents across the oceans. American and other eel larvae appear to feed only on particulate organic material referred to as marine snow (Miller et al. 2013), material produced directly from the productivity of the upper ocean, so it is possible that changes in this productivity and the quantity of food available to the larvae could influence the number of eel larvae surviving each year. Correlations have been found between atmospheric factors such as the NAO or sea surface temperature as an indicator of productivity and anguillid eel recruitment (Knights 2003; Bonhommeau et al. 2008). Other factors such as frontal location and ocean currents can also affect the success of larval eel transport from the spawning areas to the coastal habitats where they recruit (Friedland et al. 2007; Miller et al. 2009). This is not the limit of the problems facing eels, however, because introduced parasites can affect their swimbladders, viruses can infect them, and environmental contaminants can weaken them, all of these issues probably compounding to reduce the ability of migrating silver eels to reach their spawning areas successfully.

Is the American Eel Endangered?

Relative to other commercially important fish species of North America, the American eel has been largely ignored by national regulatory agencies, for many years. The US National Marine Fisheries Service published a report on the American eel (Fahay 1978) and the US Fish and Wildlife Service (USFWS) and the US Army Corps of Engineers commissioned a report to overview what was known about the species (Facey and Van Den Avyle 1987). However, it was not a priority food or sportfish for Americans, so perhaps US agencies did not pay as much attention to it as they did to high-public-profile species such as striped bass, bluefish, the declining stocks of Atlantic cod, or the endangered Atlantic salmon. That situation changed, however, when two Maine brothers became so distressed at finding silver eels killed by passage through the turbines of dams that they decided to do something about it (Prosek 2010). In the US, any citizen can submit a petition to the USFWS to propose that a species be declared endangered, so this is exactly what the two brothers did. One had worked before on several petitions to list creatures as endangered in Maine, notably the Atlantic salmon, so they were already familiar with procedures, but they still researched the subject carefully and used the data on eel declines in the St Lawrence (see Casselman et al. 1997a; Casselman 2003) to support their case. The USFWS, as was their duty under law, looked into the issue in a 90-day examination in 2005 (USFWS 2005), and they soon realized that there was insufficient information within their agency or even within any other US government agency to accept or reject the claim of the petition. They therefore initiated a year-long process called a status review, in which experts on all aspects of the species in question were to be assembled for a series of workshop-type meetings to present the existing state of knowledge about the species. For the American eel, this meant that scientific experts on the life of these eels in both freshwater and the ocean needed to be included, as well as experts on the technicalities of barriers. The goal of the process was to collect sufficient information to make an informed decision about whether or not the American eel was endangered. Although it was a serious occasion, at the end of one workshop session the question was posed whether anyone had anything else to add, especially something giving a perspective from a different side. Some additional information was then shared before one scientist who studied American eels produced a poem:

More eels
Eels in the ocean
Eels in the stream
Eels in the turbines,
I could just scream!

More eels for sushi
More eels for bass
This eel decline
Is quite a morass!

We all want more eels
Don't know who's the worst
But more eels for elvers
Have got to come first!

Paul Angermeier (from the American eel USFWS Workshop Minutes).

After the public process of gathering information from the experts, the USFWS evaluated the information and concluded that the American eel was neither threatened nor endangered (USFWS 2007). Their conclusion was presented in a short summary: "We, the US Fish and Wildlife Service (USFWS), announce our 12-month finding on a petition to list, under the Endangered Species Act of 1973 (Act), as amended, the American eel (*A. rostrata*) as a threatened or endangered species throughout its range. After a thorough review of all available scientific and commercial information, we find that listing the American eel as either threatened or endangered is not warranted at this time. We ask the public to continue to submit to us any new information that becomes available concerning the status of or threats to the species. This information will help us to monitor and encourage the ongoing conservation of this species." That statement was followed by a 31-page report on the scientific and technical details of the findings (USFWS 2007). This outcome, or at least the way the report approached the issue, was a surprise to those who had petitioned the USFWS (Prosek 2010), but with eel recruitment continuing each year with no evidence of massive disappearance of eels except in the St Lawrence River drainage, there was no scientific basis upon which to conclude otherwise. Despite that finding, another petition has been filed to the USFWS recently (CESAR 2010).

In Canada, however, a wider range of actions was already underway. In 2004, the Ontario Ministry of Natural Resources officially closed all commercial fishing for eels as a result of the dramatic decline of eels in the St Lawrence River system, the first and only jurisdiction in the world to have taken such an action for anguillid eels. In 2007, eels in Ontario were officially classified as endangered under the Ontario Endangered Species Act (MacGregor et al. 2012). The Canadian government listed the status of the American eel as being of special concern in 2006 and as being threatened in May 2012 (COSEWIC 2012), resulting in a Recovery Strategy being put into place with recommendations and approaches to increase the escapement of silver eels and recruitment (MacGregor et al. 2012). Further, in the US also in May 2012, the Atlantic States Marine Fisheries Commission (ASMFC) circulated a press release indicating that their benchmark assessment recently completed had confirmed that the American eel stock in US waters was depleted (ASMFC 2012).

In Canada too, because of their long, well-documented association with eels, Donald Marshall and the Mi'kmaq First Nations people used their experience with eels and eel fishing to argue, in the Supreme Court of Canada, that they had a treaty right to harvest marine resources for commercial purposes. They won that landmark decision (Supreme Court of Canada, R. V. Marshall 1999(3): 3ff). Then in 2008, Aboriginals assembled in Ottawa, Canada, to discuss the status of the eel and prepared an official Aboriginal People's American Eel Resolution. It called for a number of actions, including that the export fishery for glass eels be closed (i.e. glass eels were only to be used for conservation stocking) and that the American eel in Canada be considered as threatened. The indication was therefore that Aboriginals would give up many of their long-standing rights to using eels to see them restored. This resolution was signed by Aboriginal elders from across the range of eels in Canada, from the Algonquins of Ontario to the Mi'kmaq of the Maritimes (including Donald Marshall). Aboriginal concerns about the declining American eel and their long, well-documented association with the fish, which they considered to be an important reflection of their identity, led to a thought-provoking document by the Algonquins of Ontario expressing the desire to see eels returned to the large Ottawa River Basin (Algonquins of Ontario 2012), where they are now virtually absent because of the extensive construction of numerous large dams many years ago (MacGregor et al. 2012).

Much has been done since the 2003 Québec Declaration of Concern to restore the world's anguillid eels, but compared with the European eel that has been Red Listed by the International Union for Conservation of Nature (IUCN), the American eel is managed less intensively. The European eel was proposed for listing in Appendix II of CITES, and listing came into force in March 2009. In December 2010, European Union member states suspended all exports and imports of European eel commodities, probably placing further pressure on the American eel because the extensive worldwide culture of eels was left to depend almost entirely upon American glass eels and elvers, likely affecting the recovery plans for the species.

It is obvious that concerns about the status of the American eel are now considerable, and international cooperation and coordination is needed to help protect a species that has massively reduced watersheds in which to live before returning to the ocean to spawn. Therefore, the size of its population must be much smaller than it was historically. Awareness of this is a first step in learning how one may help the species survive in the ever-changing world. Understanding eels from all perspectives of our relationships with them is important in facilitating efforts to protect them and educating the general public about them. A recent book, "Eels on the Move: mysterious creatures over millions of years" by the current guest editors (Kuroki and Tsukamoto 2012) overviewed a range of subjects including the biology, life history and cultural aspects of how humanity has viewed eels and used them for food. It offers a unique view into what scientists have learned about eels and the relationship between human society and eels. As awareness of eels increases worldwide through similar books and the present one, and efforts to conserve eels intensify, the American eel will hopefully provide refreshed hope that humankind and eels can live together and flourish as they share the waters of the rivers, lakes and estuaries of eastern North America for millennia to come.

Acknowledgements Knowledge of the relationships between the American eel and humankind by many others gave the perspective needed for us to write parts of this chapter. In addition, Alex J. Haro provided some of the artwork material, and James Prosek willingly gave both useful information and artwork material.

References

Algonquins of Ontario (2012) Returning Kichisippi Pimisi—the American eel—to the Ottawa River basin. Algonquins of Ontario consultation office, Pembroke, 14 pp

ASMFC (Atlantic States Marine Fisheries Commission) (2012) ASMFC stock assessment overview: American eel, May 2012, 6 pp. www.asmfc.org

Boëtius J, Harding EF (1985) A re-examination of Johannes Schmidt's Atlantic eel investigations. Dana 4:129–162

Bonhommeau S, Chassot E, Planque B, Rivot E, Knap AH, Le Pape O (2008) Impact of climate on eel populations of the northern hemisphere. Mar Ecol Progress Ser 373:71–80

Casselman JM (2003) Dynamics of resources of the American eel, *Anguilla rostrata*: declining abundance in the 1990s. In: Aida K,

Tsukamoto K, Yamauchi K (eds) Eel biology. Springer, Tokyo, pp 255–274

Casselman JM, Cairns DK (eds) (2009) Eels at the edge: science, status, and conservation concerns. Proceedings of the 2003 international eel symposium, vol 58. American Fisheries Society Symposium, Bethesda, MA

Casselman JM, Marcogliese LA, Hodson PV (1997a) Recruitment index for the upper St. Lawrence River and Lake Ontario eel stock: a re-examination of eel passage at the R. H. Saunders hydroelectric generating station at Cornwall, Ontario, 1974–1995. In: Peterson RH (ed) The American eel in eastern Canada: stock status and management strategies. Proceedings of the eel workshop, Québec City, 13–14 January 1997, Can Tech Rep Fish Aquat Sci 2196:161–169

Casselman JM, Marcogliese LA, Stewart T, Hodson PV (1997b) Status of the upper St. Lawrence River and Lake Ontario American eel stock—1996. In: Peterson RH (ed) The American eel in eastern Canada: stock status and management strategies. Proceedings of the eel workshop, Québec City, 13–14 January 1997, Can Tech Rep Fish Aquat Sci 2196:106–120

Castonguay M, Hodson PV, Couillard CM, Eckersley MJ, Dutil JD, Verreault G (1994) Why is recruitment of the American eel, *Anguilla rostrata*, declining in the St Lawrence River and Gulf? Can J Fish Aquat Sci 51:479–488

CESAR (Council for Endangered Species Act Reliability) (2010) Petition to list American eel (*Anguilla rostrata*) as a threatened species under the endangered species act. www.fws.gov/northeast/newsroom/pdf/American_eel_petition_100430.pdf

Chow S, Kurogi H, Mochioka N, Kaji S, Okazaki M, Tsukamoto K (2009) Discovery of mature freshwater eels in the open ocean. Fish Sci 75:257–259

COSEWIC (Committee on the Status of Endangered Wildlife in Canada) (2012) COSEWIC assessment and status report on the American eel *Anguilla rostrata* in Canada. Committee on the status of endangered wildlife in Canada, Ottawa. xii+, 109 pp. www.registrelep-sararegistry.gc.ca/default_e.cfm

Dekker W (1998) Long-term trends in the glass eels immigrating at Den Oever, the Netherlands. Bull Fr Peche Piscic 349:199–214

Dekker W, Casselman JM, Cairns DK, Tsukamoto K, Jellyman D, Lickers H (2003) Worldwide decline of eel resources necessitates immediate action: Québec declaration of concern. Fisheries 28:2830

du Creux E (1664) History of Canada or New France, vol 1. The Champlain Society, Toronto

Facey DE, Van Den Avyle MJ (1987) Species profiles: life histories and environmental requirements of coastal fishes and invertebrates (North Atlantic): American eel. Biological report 82(11.74). Biological report US fish and wildlife service 82(11.74). US Army Corps of Engineers, TR EL-82-4, 1987, 28 pp

Fahay MP (1978) Biological and fisheries data on American eel, *Anguilla rostrata* (LeSueur). US Department of Commerce, national marine service technical report, vol 17, Northeast Fisheries Center, Highlands, 82 pp

Friedland KD, Miller MJ, Knights B (2007) Oceanic changes in the Sargasso Sea and declines in recruitment of the European eel. ICES J Mar Sci 64:519–530

Haro A (2003) Downstream migration of silver phase anguillid eels. In: Aida K, Tsukamoto K, Yamauchi K (eds) Eel biology. Springer, Tokyo, pp 215–222

Haro AJ, Castro-Santos T (2000) Behavior and passage of silver phase American eels, *Anguilla rostrata* (LeSueur), at a small hydroelectric facility. Dana 12:33–42

Haro AJ, Krueger WH (1988) Pigmentation, size, and migration of elvers (*Anguilla rostrata* (Lesueur)) in a coastal Rhode-Island stream. Can J Zool 66:2528–2533

Haro A, Richkus W, Whalen K, Hoar A, Busch W-D, Lary S, Brush T et al (2000) Population decline of the American eel: implications for research and management. Fisheries 25:7–16

Helfman GS, Clark JB (1986) Rotational feeding: overcoming gape-limited foraging in anguillid eels. Copeia 1986:679–685

Helfman GS, Facey DE, Hales S, Bozeman EL (1987) Reproductive ecology of the American eel. In: Dadswell MJ, Klauda RJ, Moffitt CM, Saunders RL, Rulifson RA, Cooper JE (eds) Common strategies of anadromous and catadromous fishes, vol 1. American Fisheries Society Symposium, Bethesda, pp 42–56

Hjort J (1910) Eel larvae from the central Atlantic. Nature 85:104–106

Hurrell JW (2005) Decadal trends in the North Atlantic oscillation: regional temperatures and precipitations. Science 269:676–679

Jessop BM (2010) Geographic effects on American eel (*Anguilla rostrata*) life history characteristics and strategies. Can J Fish Aquat Sci 67:326–346

Jessop BM, Cairns DK, Thibault I, Tzeng WN (2008) Life history of American eel *Anguilla rostrata*: new insights from otolith microchemistry. Aquat Biol 1:205–216

Junker-Andersen C (1988) The eel fisheries of the St Lawrence Iroquoians. North Am Archaeol 9:97–121

Kleckner RC, McCleave JD (1985) Spatial and temporal distribution of American eel larvae in relation to North Atlantic Ocean current systems. Dana 4:67–92

Kleckner RC, McCleave JD (1988) The northern limit of spawning by Atlantic eels (*Anguilla* spp.) in the Sargasso Sea in relation to thermal fronts and surface water masses. J Mar Res 46:647–667

Kleckner RC, McCleave JD, Wippelhauser GS (1983) Spawning of American eel, *Anguilla rostrata*, relative to thermal fronts in the Sargasso Sea. Environ Biol Fish 9:289–293

Knights B (2003) A review of the possible impacts of long-term oceanic and climate changes and fishing mortality on recruitment of anguillid eels of the northern hemisphere. Sci Total Environ 310:237–244

Krueger WH, Oliveira K (1999) Evidence for environmental sex determination in the American eel. Environ Biol Fish 55:381–389

Kuroki M, Tsukamoto K (2012) Eels on the move: mysterious creatures over millions of years. Tokai University Press, Tokyo, 278 pp

MacGregor R, Casselman J, Greig L, Dettmers J, Allen WA, McDermott L, Haxton T (2012) Draft recovery strategy for the American eel (*Anguilla rostrata*) in Ontario. Ontario recovery strategy series. Prepared for Ontario Ministry of Natural Resources, Peterborough, Ontario, x +pp 128+ Appendices

McCleave JD (2001) Simulation of the impact of dams and fishing weirs on reproductive potential of silver-phase American eels in the Kennebec River Basin, Maine. North Am J Fish Manag 21: 592–605

McCleave JD (2003) Spawning areas of the Atlantic eels. In: Aida K, Tsukamoto K, Yamauchi K (eds) Eel biology. Springer, Japan, pp 141–155

McCleave JD, Kleckner RC (1982) Selective tidal stream transport in the estuarine migration of glass eels of the American eel (*Anguilla rostrata*). J Cons Int Explor Mer 40:262–271

McCleave JD, Kleckner RC, Castonguay M (1987) Reproductive sympatry of American and European eels and implications for migration and taxonomy. In: Dadswell MJ, Klauda RJ, Moffitt CM, Saunders RL, Rulifson RA, Cooper JE (eds) Common strategies of anadromous and catadromous fishes, vol 1. American Fisheries Society Symposium, Bethesda, pp 286–297

Miller MJ, McCleave JD (1994) Species assemblages of leptocephali in the subtropical convergence zone of the Sargasso Sea. J Mar Res 52:743–772

Miller MJ, Kimura S, Friedland KD, Knights B, Kim H, Jellyman DJ, Tsukamoto K (2009) Review of ocean-atmospheric factors in the Atlantic and Pacific oceans influencing spawning and recruitment of anguillid eels. In: Haro AJ, Smith KL, Rulifson RA, Moffitt CM, Klauda RJ, Dadswell MJ, Cunjak RA et al (eds) Challenges for diadromous fishes in a dynamic global environment, vol 69. American Fisheries Society Symposium, Bethesda, pp 231–249

Miller MJ, Chikaraishi Y, Ogawa NO, Yamada Y, Tsukamoto K, Ohkouchi N (2013) A low trophic position of Japanese eel larvae indicates feeding on marine snow. Biol Lett 9:20120826

Munk P, Hansen MM, Maes GE, Nielsen TG, Castonguay M, Riemann L, Sparholt H et al (2010) Oceanic fronts in the Sargasso Sea control the early life and drift of Atlantic eels. Proc R Soc Ser B 277: 3593–3599

Oliveira K (1999) Life history characteristics and strategies of the American eel, *Anguilla rostrata*. Can J Fish Aquat Sci 56:795–802

Oliveira K, Hable WE (2010) Artificial maturation, fertilization, and early development of the American eel (*Anguilla rostrata*). Can J Zool 88:1121–1128

Oliveira K, McCleave JD (2000) Variation in population and life history traits of the American eel, *Anguilla rostrata*, in four rivers in Maine. Environ Biol Fish 59:141–151

Orr RB (1917) Ontario Indians: their fish, fisheries, and fishing appliances. 29th annual archaeological report. King's Printer, Toronto, pp 24–43

Parker SJ (1995) Homing ability and home range of yellow-phase American eels in a tidally dominated estuary. J Mar Biol Assoc UK 75:127–140

Pennisi E (1991) Gone eeling: luck and science face off in two eel-seeking adventures. Sci News 140:297–299

Prosek J (2010) Eels: an exploration, from New Zealand to the Sargasso, of the world's most mysterious fish. Harper–Collins, New York, p 287

Richkus WA, Dixon DA (2003) Review of research and technologies on passage and protection of downstream migrating catadromous eels at hydroelectric facilities. In: Dixon DA (ed) Biology, management, and protection of catadromous eels, vol 33. American Fisheries Society Symposium, Bethesda, pp 377–388

Schmidt J (1922) The breeding places of the eel. Phil Trans R Soc B 211:179–208

Schmidt J (1927) Eel larvae in the Faroe channel. J Cons Int Explor Mer 2:38–43

Schmidt J (1935) Danish eel investigations during 25 years (1905–1930). The Carlsberg Foundation, Copenhagen, 16 pp

Schoth M, Tesch F-W (1982) Spatial distribution of 0-group eel larvae (*Anguilla* sp.) in the Sargasso Sea. Helgoländer Meeresun 35: 309–320

Shinoda A, Aoyama J, Miller MJ, Otake T, Mochioka N, Watanabe S, Minegishi Y et al (2011) Evaluation of the larval distribution and migration of the Japanese eel in the western North Pacific. Rev Fish Biol Fish 21:591–611

Tesch F-W (2003) The eel: biology and management of anguillid eels. Blackwell, London

Tesch F-W, Wegner G (1990) The distribution of small larvae of *Anguilla* sp. related to hydrographic conditions 1981 between Bermuda and Puerto Rico. Int Rev Hydrobol Hydrogr 75: 845–858

Thwaites RG (1903) A new discovery of a vast country in America, by Father Louis Hennepin. A. C. McClurg and Co, Chicago

Tsukamoto K (1992) Discovery of the spawning area for Japanese eel. Nature 356:789–791

Tsukamoto K, Aoyama J, Miller MJ (2009) The present status of the Japanese eel: resources and recent research. In: Casselman JM, Cairns DK (eds) Eels at the edge: science, status, and conservation concerns, vol 58. American Fisheries Society Symposium, Bethesda, pp 21–35

Tsukamoto K, Chow S, Otake T, Kurogi H, Mochioka N, Miller MJ, Aoyama J et al (2011) Oceanic spawning ecology of freshwater eels in the western North Pacific. Nat Commun 2:179

Tsukamoto K, Mochioka N, Miller MJ, Koyama S, Watanabe S, Aoyama J (2013) Video observation of an eel in the *Anguilla japonica* spawning area along the West Mariana Ridge. Fish Sci 79:407–416

USFWS (United States Fish and Wildlife Service) (2005) Endangered and threatened wildlife and plants; 90-day finding on a petition to list the American eel as threatened or endangered. Fed Reg 70(128):38849–38861

USFWS (United States Fish and Wildlife Service) (2007) Endangered and threatened wildlife and plants; 12-month finding on a petition to list the American eel as threatened or endangered. Fed Reg 72(22):4967–4997, Available at www.archives.gov

Verreault G, Pettigrew P, Tardif R, Pouliot G (2003) The exploitation of the migrating silver American eel in the St. Lawrence River estuary, Québec, Canada. In: Dixon DA (ed) Biology, management, and protection of catadromous eels, vol 33. American Fisheries Society Symposium, Bethesda, pp 225–234

Wippelhauser GS, McCleave JD (1988) Rhythmic activity of migrating juvenile American eels (*Anguilla rostrata* LeSueur). J Mar Biol Assoc UK 68:81–91

Wirth T, Bernatchez L (2003) Decline of North Atlantic eels: a fatal synergy? Proc R Soc London B 270:681–688

Epilogue

This is the first book to deal with all aspects of eels in a single volume. In it, there are descriptions of almost all that is known about eels in various countries, focusing particularly on social and cultural aspects. From a social perspective, we have addressed aspects such as fishing, distribution, economics, food culture, aquaculture and problems of resources and environments; culturally, we have delved into idioms, arts, traditions, legends, mythology, archaeology and memorial services (see Fig. 1 in the Preface).

In terms of social sciences, we have seen eels being exploited through the vicissitudes of human society, and how the waters have been muddied by issues of world trade and economics. Culturally, we have shown how affection and respect have been fermented through the long relationship of familiarity between humans and eels, and how this has been transformed into a rich, warm-hearted culture.

The objectives of this book when we started compiling it were to compare each of these eel-related issues laterally among the countries involved, and thereby to obtain a comprehensive understanding of the total relationship between eels and humans in the world. We believe that such an understanding of the tight relationship between eels and humankind will increase awareness of the importance of eels on earth, and enhance feelings of affection and respect towards them.

How then should one relate to eels in future, although "relate" is not quite the right word. A relationship should be two-way, but that between humans and eels has always been wholly one-sided. From the perspective of the eels, the relationship must surely have been hugely burdensome, and thanks to it, eel stocks have declined dramatically all around the world. Halting this decline and restoring the original balance is the responsibility of we humans who imposed such a one-sided relationship on eels.

Eels are distributed across many parts of the world, across the borders of individual countries and continents, so to protect them, we need international cooperation based on solid relationships of trust. That will be difficult to achieve, of course, without collaborative efforts and intergovernmental understanding between industry, the public sector and academia.

Meanwhile, the eel species of particular use to man should be "domesticated," much as cows, pigs and chickens are, so that we can manage all the processes from eggs to adult fish through human endeavour. Seedlings for aquaculture are required in large volumes and are always in a state of shortage, so we need urgently perfect the technology for producing artificial seedlings to meet even a part of this huge demand with artificial glass eels raised from eggs. By so doing, we would hopefully reduce the overexploitation of eels in the wild and take steps to conserve certain species.

After editing this book, we feel that we may already have part-succeeded in reaching our goal, by providing an all-embracing overview that gathers extensive knowledge on eels in a single volume. We thank the contributors of all the chapters for introducing valuable knowledge on eels from various fields, gleaned sometimes in the course of writing the various contributions. In addition, we hope that our readers will discover a new side to eels and broaden their outlook on this creature. We would be glad too if this could help foster deeper affection for eels in people's minds, and ultimately lead to the conservation of all species of eel. Finally we fervently hope, with all sincerity, that this mysterious creature called the eel will remain close to us for ever.

Katsumi Tsukamoto
Mari Kuroki
12 December 2012

K. Tsukamoto and M. Kuroki (eds.), *Eels and Humans*, Humanity and the Sea,
DOI 10.1007/978-4-431-54529-3, © Springer Japan 2014

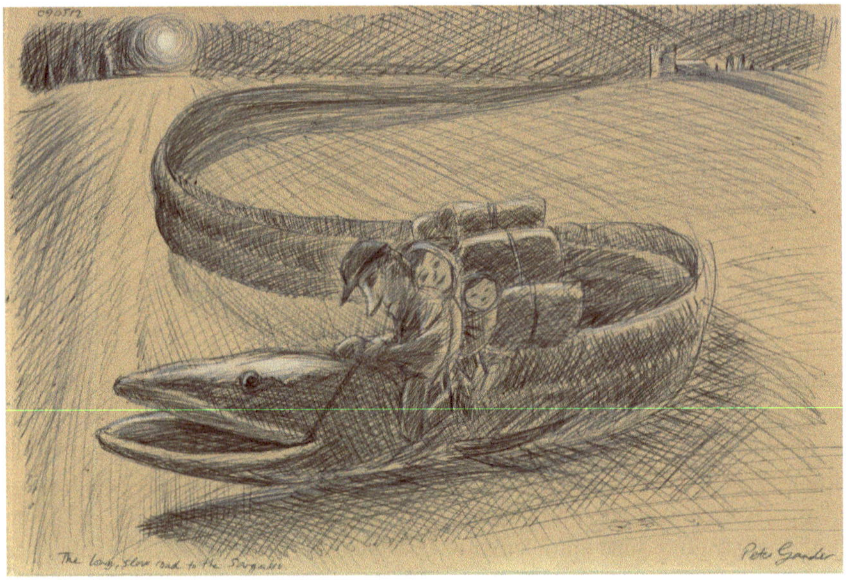

The long slow road to the Sargasso Sea, a drawing in pencil and ballpoint pen (copyright Peter Gander)

Index

K. Tsukamoto and M. Kuroki (eds.), *Eels and Humans*, Humanity and the Sea,
DOI 10.1007/978-4-431-54529-3, © Springer Japan 2014

R

S

T

U

V

MIX
Papier aus verantwortungsvollen Quellen
Paper from responsible sources
FSC® C105338

If you have any concerns about our products,
you can contact us on
ProductSafety@springernature.com

In case Publisher is established outside the EU,
the EU authorized representative is:
Springer Nature Customer Service Center GmbH
Europaplatz 3, 69115 Heidelberg, Germany

Printed by Libri Plureos GmbH
in Hamburg, Germany